Bell Labs
Life in the Crown Jewel

Bell Labs

Life in the Crown Jewel

Narain Gehani

SILICON PRESS
Summit, NJ 07901, USA
www.silicon-press.com

**Silicon Press
25 Beverly Road
Summit, NJ 07901
USA**

First Edition
Printing 9 8 7 6 5 4 3 2 1   Year 07 06 05 04 03

Printed on acid-free paper.

Library of Congress Cataloging-in-Publication Data

Gehani, Narain,
  Bell Labs : life in the crown jewel / Narain Gehani.
      p. cm.
Includes bibliographical references and index.
  ISBN 0-929306-27-9
  1. Bell Labs Innovations. 2. Electric engineering--Research--United
States. 3. Corporate culture. I. Title.
TK415.B45 G44 2002
621.3'072073--dc21

                                                              2002012730

Contents

Preface

WANT TO WORK with Nobel Prize winners? Or with the folks who created the UNIX™ system or designed the C and C++ programming languages, or those who played a key role in creating object database technology? Or follow in the footsteps of the inventors of the transistor or the Karmarkar algorithm or those who did pioneering work in movies with sound, radar, and lasers? Or collaborate with world famous scientists?

Bell Labs, the research part of Bell Labs to be precise, offered such an environment with a wealth of fame, expertise, and history. This legendary institution, the greatest research lab of the twentieth century, was the "in place" to work for researchers from not only the USA, but also all around the world. Budding scientists, and even established ones, aspired to join Bell Labs for its academic freedom, its resources, and the opportunity to do world-class research, possibly in collaboration with world famous colleagues.

Bell Labs has provided a great environment for research and advancing science. In this fertile environment, Bell Labs researchers have been extremely prolific in producing fabulous inventions and new ideas. Their contributions have been recognized with numerous prestigious awards such as Nobel Prizes and National Medals of Science.

From 1925 until 1984, Bell Labs was a separate company jointly owned by AT&T and its subsidiary Western Electric. After the AT&T divestiture in 1984, Bell Labs became a division of AT&T. Following the next breakup of AT&T, in 1996, Bell Labs became a division of Lucent. The 1984 breakup catapulted AT&T into the competitive arena, which put pressure on Bell Labs to steer away from basic research towards industrial research.

I had the privilege and honor of working at this legendary institution for twenty-three years, most of them wonderful and productive. I joined in 1978 as a member of technical staff. Before this, I had been an assistant professor of computer science at the State University of New York at Buffalo, NY. My technical expertise is in software in its various flavors – Web systems, mobility, programming languages, parallel programming, databases, and so forth. I started at Bell Labs by spending my first year in a Bell Labs organization that was developing a UNIX platform known as the Programmer's Workbench. Bell Labs was a huge organization in those days, but only a small part of it was involved in basic research. About a year after joining, I transferred to the research part of Bell Labs. Over the years, I was instrumental in creating several innovative software systems, wrote many technical papers, and filed and was awarded many patents.

In 1985, I was made a distinguished member of technical staff. In 1993, I was appointed to head the newly created Database Systems Research Department. Then in December 2000, I was appointed research vice president, Bell Labs Research Silicon Valley (BLRSV). Despite the Silicon Valley in my job title, I also had organizational responsibility for the Database Systems Research Department in New Jersey. After the closure of the BLRSV, I was asked to take on the additional responsibility of managing the research department in Naperville, IL and my title morphed to research vice president, Communications Software Research. Within Bell Labs, this organization was also known as center 1138.

For about a year during my tenure as the head of the Database Systems Research Department, I was also the president of Maps On Us (www.mapsonus.com), a Web service that provides maps, routes, and directions. I retired from Bell Labs in the summer of 2001.

My goal in writing this book is to tell you about Bell Labs, some of its accomplishments and what it was like to work at this fascinating institution, the greatest research lab of the twentieth century. In the case of some anecdotes, to preserve anonymity, I have changed details such as names without affecting the stories. I will also talk about the challenges facing Bell Labs as it moves from basic research to applied research. When I talk about Bell Labs, I mean the research part of Bell Labs, Bell Labs Research, the subject of this book.

Almost from the time I joined Bell Labs in 1978, Bell Labs was in the process of increasing software research and reducing research in other areas such as physical sciences. By the end of 2000, a majority of Bell Labs researchers were working on next generation software and applications.[1] As I was focused on software research, I had very little interaction with researchers in physical sciences. I would therefore like to emphasize that my understanding of Bell Labs and its contributions to its owners, first AT&T and then Lucent, is from the perspective of the non-physical sciences part of Bell Labs.

Narain Gehani

Acknowledgments

I AM GRATEFUL to the many persons who have helped me in a variety of ways in writing this book. Robbie Clipper Sethi's suggestions helped improve the book significantly. The discussions that I had with Alan Feuer and Charles Wetherell were very valuable. Alan, Stow Lovejoy, and Anoo Verghis' comments helped me organize the manuscript better and improve it. Meenakshi Hirani's thoughts helped shape the book significantly. Bill Roome's historical knowledge about Bell Labs helped make the book more accurate. Bob Cmelik made many suggestions that helped make the book better. Indu Gehani proofread the manuscript. Robi and Josh Weinreich encouraged me to write the book. Viju Verghis and Shelley Beckler Modi gave thumbs up. Dean Polnerow, president and founder of Switchboard Inc., was instrumental in getting me permission to publish Maps On Us images.

Bell Labs provided a wonderful work environment. Susan Witzel made it possible for me, in the last few years, to focus on the challenging aspects of my job. With Vinod Anupam, Bob Cmelik, Jacques Gava, Dan Lieuwen, and Bill Roome, I was involved in building important and innovative computer systems. These researchers put their heart and soul into their work, working long and hard, just for the pure pleasure of accomplishing something.

Finally, I appreciate Neel and Varun's patience with me during the long process of writing this book and their periodic checks to make sure that I was indeed making progress.

1 I Have A Job For Life!

August 15, 1978

Unless you do something criminal, like rape one of the mail girls, you have a job for life!

THIS WAS ONE of the remarks made by a manager while welcoming me to Bell Labs. I was taken aback. I was joining Bell Labs to do research. Why was he talking about rape and criminal actions? The manager had a wisp of a smile on his face, so perhaps this was his peculiar sense of humor.

Nevertheless, being risk averse I was relieved that I would never have to look for another job. Job security was an important reason for my joining Bell Labs as opposed to joining companies that downsized at every economic downturn. I was also attracted to Bell Labs by its worldwide reputation, famous people, and resources, and the opportunities it provided for innovation while working with world-class scientists.

22 Years Later

Lucent, the current corporate parent of Bell Labs, has not been doing well. There is tremendous negative press about Lucent. There has even been talk about Lucent going bankrupt. Starting in early 2000, Lucent's financial results in several successive quarters were worse than expected and/or contained bad news. Revenues were spiraling downwards with costs seemingly out of control. Reflecting the state of Lucent, its stock kept on taking a beating.

Chairman and CEO Rich McGinn was fired on October 23, 2000 and replaced by his predecessor, Henry Schacht, who took over as the interim chairman and CEO. The morale went up because Schacht was respected by the Lucent folks. He had been Lucent's first CEO. Schacht soon put a restructuring plan into place. The Bell Labs budget, along with those of other Lucent divisions was reduced.

All this had a sobering effect on the Bell Labs researchers. Many of us had worked hard developing new technologies that we thought would help the Lucent business.

February 6, 2001

February 6 was a red-letter day for me. Lucent's financial condition and the changing marketplace forced Bell Labs to throw away the "job for life" culture that was highly valued by its risk-averse researchers such as me.

On Tuesday, February 6, 2001, Bell Labs closed its Silicon Valley research facility laying off 18 researchers! The Silicon Valley facility had been opened in July 1998 to serve as a technology vanguard for the increasing number of Lucent business operations in California (as a result of acquisitions).

This event marked a big change in Bell Labs culture – from taking care of its employees with lifetime employment to hiring and firing as dictated by market conditions. To my knowledge, Bell Labs had never previously fired researchers or closed a research facility for economic reasons. Bell Labs would have liked to keep the closing as quiet as possible since it would have a negative impact on its image and would make future hiring more difficult. Thus, no official announcement was made about closing the research facility in Silicon Valley, even within Bell Labs. The closing would have gone unnoticed by the rest of the world but for the ever-vigilant *The New York Times*. The paper reported the closing next day in an article[2] titled

Lucent Closes Silicon Valley Laboratory

Quoted in this article is David Nagel, then head of AT&T Labs:

> *I think this is unprecedented, I cannot ever remember laying off an entire research group in the history of Bell Labs.*[3]

Late in the previous week, Bell Labs Research Silicon Valley (BLRSV) researchers had been asked to assemble in their conference room early Tuesday morning for an important announcement. Bill Brinkman, vice president of research, would address them, via a video link, from Lucent headquarters in Murray Hill, NJ.

BLRSV was rife with rumors about being closed. The staff assembled in the conference room that early Tuesday morning was depressed but ready to accept a closure and move on. The rumors proved to be correct. At the appointed time, Brinkman appeared on the TV screen and, without much ado, he announced the closure of BLRSV due to Lucent's difficult financial condition.

This was a very sad day for me. Several weeks earlier, I had been appointed research vice president, Bell Labs Research Silicon Valley. Besides Silicon Valley, I also had research staff in New Jersey. I was to commute to California every other week.

I had visited BLRSV for the first time in late December 2000. BLRSV had been leaderless for a while and, understandably, the morale was low. The BLRSV researchers wanted to know why the folks back in Murray Hill had left them leaderless for several months. Did Brinkman and his boss Arun Netravali, the Bell Labs president, not care about BLRSV?

Returning to New Jersey, I reported in detail the state of BLRSV to my boss, Brinkman. My report confirmed what Brinkman had been suspecting all along – Silicon Valley Labs needed serious attention. Brinkman said that we should continue to operate BLRSV. I was to try to "fix" the lab and he would review its state after one year.

Just before my next visit in late January (delayed because of a previous commitment), Brinkman asked me to meet with him. In the last few weeks, Lucent's financial condition had deteriorated, as a result of which Brinkman said that he would have to close BLRSV. I was shocked to hear this. After some discussion, I concurred with his decision. It was obvious to me that this had been a difficult decision for Brinkman.

As I was the head of BLRSV, I was present in the conference room in our Palo Alto facility where the BLRSV staff had assembled for Brinkman's announcement. I was sick to my stomach in anticipation of the closure and was worried about the possible reactions of the staff whose lives were going to be changed. I had never laid off a person in my whole career. However, the staff took the closure in stride. They listened quietly to Brinkman speak from Murray Hill and did not have many questions for him. After the announcement, they went back to their offices and started packing up. Later that day, some of the researchers told me that in Silicon Valley layoffs were quite frequent and that they had been expecting something like this. They were appreciative of the generous severance package that Lucent was offering them. Even in a financially difficult time, Lucent was trying to do the best it could do financially for the employees that it was letting go!

The closing of BLRSV was my baptism by fire into senior management.

May/June/Early July 2001

After the closure of Bell Labs Research Silicon Valley, I was asked to take on the additional responsibility of managing the Bell Labs research outpost located in Naperville, IL, which consisted of about 20 researchers – in addition to my NJ research staff. Moreover, my title changed to research vice president, Communication Software Research.

Meanwhile, Lucent's financial condition had continued to deteriorate. In an attempt to put Lucent on an even keel, Lucent chairman and CEO Henry Schacht announced a restructuring plan that called for the downsizing of Lucent. Contract employees were to be terminated and a 5+5 Voluntary Retirement Plan (VRP) was announced. According to Lucent's retirement rules, if an employee meets the two conditions

$Age \geq 50$

$Age + Number\ of\ Years\ at\ Lucent \geq 75$

then an employee is eligible to retire with a "full" pension, which the employee starts getting immediately.

The amount of the full pension is based on a formula, which computes an employee's pension based on the number of years the employee has worked with the company, an "average" salary as defined by Lucent rules, the kind of spousal coverage desired, a multiplier determined by Lucent, and so forth.

The voluntary retirement program "sweetened" the normal retirement rules to encourage employees to retire. The 5+5 in the 5+5 VRP designated the sweetening. The first 5 increased the employee's age for qualifying for retirement, thus lowering the minimum age for retirement from 50 years to 45. The second 5 in the 5+5 VRP increased the number of years an employee had worked at Lucent for the purpose of computing the employee's pension, thus increasing the pension. The 5+5 thus made it easier to qualify for retirement and gave the qualifying employees a higher pension.

The 5+5 VRP was announced on June 10 and eligible staff members had until July 11 to decide whether they would take it. Four of my staff members were eligible to take the offer. The offices of two of the researchers, Bob Arlein and Bill Roome, who were key personnel, were down the hall from my office. Their intentions to take the 5+5 VRP fluctuated with the day and, particularly in the case of Bob, the prospects of finding another job. So we drew thermometers that Bob and Bill were supposed to update when their chances of staying or going changed:

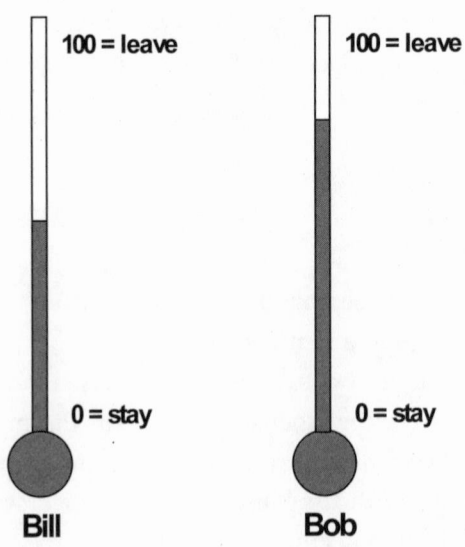

Anyone could pop into Bob and Bill's offices and see the odds of their staying. I wanted them to stay and every time I looked at the thermometers in their offices, I hoped to see the "temperatures"

lower than the last time I checked. Bob was more inclined than Bill to take the 5+5 VRP offer and leave. But in the end, both Bob and Bill decided to stay. The deteriorating job market helped them make the decision. Although their staying did not help Lucent in its downsizing effort, both Bob and Bill bring tremendous value to Lucent, as they are system builders par excellence.

The 5+5 VRP had been offered only to middle managers and below. I told my assistant, Susan Witzel, that I was relieved that the offer did not include me because I would have had a tough time deciding whether or not to take it.

Friday, July 13, 2001

Around noon, a colleague, who was also a research vice president, stopped by to chat. He had been given the choice of either taking an incentive package and leaving Bell Labs or taking on lesser responsibility. The new job would have the same base pay as his current job. He asked me to give him some advice. I said that this was a difficult decision and depended a lot on personal circumstances. If I were in his shoes, I would take the incentive package and leave. I was sure that, given my accomplishments, I would be able to find another job. From my perspective, if staying at Bell Labs meant taking a job with reduced responsibility, then the scales were tipped in favor of leaving. After some deliberation, my colleague decided to stay.

I did not think I was going to be affected by the continuing downsizing. Software was important for Lucent, and Netravali had told me software would not be affected by the downsizing. However, I was worried about the impact of the downsizing on the researchers. But I had a meeting to attend which made me push aside the thoughts about downsizing so that I could deal with the issues at hand. Late in the afternoon, I returned to my office from a meeting and saw on the telephone "call log" that Brinkman had tried to reach me but he had left no message.

Given the current downsizing environment in Bell Labs and in Lucent, many thoughts crossed my mind and I started to prepare for the worst.

Monday, July 16, 2001

As I was on my way to work in the morning, my assistant Susan Witzel called me on my cell phone and said that Brinkman wanted to meet with me. Now I was really concerned. About what, I asked. She did not know, so I asked her to find out. She checked with Brinkman's assistant and told me that Brinkman wanted to talk to me about a transfer that I had approved.

The transfer was routine, but maybe Brinkman had some questions about it. Nevertheless, given Lucent's financial condition, I was concerned. The meeting with Brinkman was scheduled for later in the day at 1:30 PM.

On entering his office, I asked Brinkman if I should be looking for a job. Brinkman said that was not the case. However, he said that they were closing the Naperville, IL research department, which represented about half my staff. He said that other remote research departments in Greece, England, and India were also going to be closed. The exception was China, where Bell Labs Research's presence was a Lucent business necessity.

Since about half my staff would be leaving, I was affected. I could take an incentive package, or I could take a job with less responsibility but with the same pay. I was of course dismayed to learn all this, but without any hesitation, I told him that I would take the incentive package. (At the same time I also qualified for early retirement.) To my surprise, Brinkman seemed surprised. He said that before I made my decision I should talk to Netravali. Even though I would talk with Netravali, I knew that I would leave.

August 15, 2001

I left Bell Labs after exactly 23 years. I was not quite sure what I would do next and was somewhat apprehensive. However, I was not unhappy about leaving Bell Labs. I was leaving a Bell Labs that was very different from the one I had joined many years ago, one which was in the process of being transformed into an industrial research lab, and one which would be facing serious challenges along with its parent, Lucent. I had been at Bell Labs a long time. It was time to try something different.

When I had joined Bell Labs, it was part of the mega-giant AT&T, which at the time had a payroll of about one million employees. AT&T was then a dominant corporation on the American scene, a trusted company that was affectionately known as Ma Bell. AT&T provided telephone services to most people in America. It wielded tremendous political and social clout all over the country because it provided a vital service and, more importantly, because its employees and retirees could be found in almost every community.

When I left Bell Labs in August 2001, it was owned by Lucent, a company that had shrunk to about 70,000 employees and which by now did not have clout anywhere close to that of the old Ma Bell. In the spring of 2002, Lucent announced that, within a few months, it would cut its workforce to 45,000. Lucent then announced more financial bad news in September 2002 and hinted at further job cuts. In October 2002, Lucent announced even more bad news and 10,000 job cuts that would reduce its workforce to 35,000. One analyst expected Lucent's workforce to go down to 30,000!

The telecommunications world changed very quickly for Lucent. From an insatiable demand for telecommunications equipment, such as routers of all kinds and optical and wireless equipment, within the space of one year (2000), the telecommunications industry found itself saddled with excess capacity. Telecommunications service providers now had much more capacity than they could sell. The excess capacity was caused by over investment and technological advances, such as those that dramatically increased the data that could be pumped through existing optical fibers. Moreover, telecommunications traffic was not growing at the rate expected by the service providers. This meant lower revenues, which were further lowered by the tremendous price pressures caused by intense competition. The stronger service providers reduced their purchases of telecommunications equipment while the weaker ones collapsed. As a result, sales of telecommunications equipment fell dramatically, badly hurting telecommunications equipment makers such as Lucent. In addition, several of them, including Lucent, were further hurt financially because, when their customers went bankrupt, they were forced to write off hundreds of millions of dollars in loans. They had financed the sales of their equipment to these customers, a scheme known as "vendor financing."

The telecommunications sector quickly went from being the darling of Wall Street to a sector that should be avoided, an untouchable. Lucent was even the subject of rumors of bankruptcy, which it was forced to deny with a press release:

> *The rumors that Lucent is filing for bankruptcy are baseless and irresponsible.*[4]

Despite the difficulties Lucent has had in the first few years of the new millennium, some analysts think that Lucent will weather the downturn in the telecommunications industry while many of its weaker rivals will go under as some have already done.

Postscript

Leaving Bell Labs left me with a strange feeling, a feeling that I find hard to describe. I used to imagine myself wandering the halls of Murray Hill many years into the future. Although I can still imagine doing that, it is not likely that I will now be doing so. I have good memories of the many wonderful years I spent at Murray Hill, working with interesting people, passionate researchers, famous colleagues, and more. Together, we explored many ideas, published papers, built innovative systems, played backgammon and bocce, saw movies, and so forth. I got the chance to do many interesting projects. The one that stands out is Maps On Us (www.mapsonus.com), which Bill Roome and I created. Every day, tens of thousands of people use Maps On Us, years after we created it!

2   The Crown Jewel

MURRAY HILL, NJ

THE FAMED BELL Labs Research is located in a sprawling building complex, which houses about 4000 employees, in Murray Hill, NJ. The complex is garlanded by beautifully manicured lawns up in the front and woods in the back. It consists of many adjoining buildings, each with long corridors, some of them with overhead emergency showers (to address chemical emergencies) from yesteryear. The complex is a labyrinth and it takes some familiarity with it to find one's way.

Visitors to Bell Labs see a magnificent facade behind which is an imposing atrium with a high ceiling. On the right of the reception desk is a bust of Alexander Graham Bell, the inventor of the telephone and the founder of AT&T. The bust is not directly in front of the visitor walking in and, as a result, it often goes unnoticed. Only visitors with a bit of time on their hands, such as visitors waiting for their host to escort them, may happen to wander over to the bust and read the following words etched in its base:

> *Leave the beaten track occasionally and dive into the woods. You will be certain to find something you have never seen before.*

Alexander Graham Bell's words beautifully describe what research is all about. Bell Labs researchers, following founder Bell's advice, have discovered and invented prolifically, as no other group of scientists has ever done before, in the over three quarters of a century since Bell Labs was founded.

Bell Labs has made the name Murray Hill famous all over the world as the home of the greatest research lab of the twentieth century. People are surprised when they learn that there is no city or town named Murray Hill in New Jersey. Murray Hill, NJ is actually the name of an unincorporated area split between the tax domains of two jurisdictions, the town of Berkeley Heights and the borough of New Providence. Bell Labs is itself partly in Berkeley Heights and partly in New Providence.

Murray Hill was the headquarters of Bell Labs until its status changed from that of an independent company to a division of AT&T as a result of the 1984 AT&T breakup. Following this change, Bell Labs senior management continued to be located primarily at Murray Hill, and it continued as the primary location of Bell Labs Research. In the 1980s, Holmdel was the de facto Bell Labs headquarters because Ian Ross, the sixth Bell Labs president, had his offices in Holmdel. The Murray Hill complex is now the headquarters of Lucent Technologies, the current corporate parent of Bell Labs.

Bell Labs consists of more than just the research organization, although the number of other organizations is now down to a handful. When I joined in 1978 most of Bell Labs consisted of development organizations. In addition to the Murray Hill facility, there were numerous other Bell Labs locations in New Jersey, the more famous ones being Holmdel and Whippany. There was also a big, but not as famous, facility in Piscataway. There were Bell Labs facilities in many other states, but the big ones were in Illinois, Colorado, and Massachusetts. With the exception of Holmdel, these other Bell Labs locations housed only development organizations. In the case of Holmdel, a small fraction of Research was co-located with the development organizations. The development organizations were moved out of Bell Labs in 1984 following the AT&T breakup.

When the world thinks of Bell Labs, they are thinking of its famous research organization, Bell Labs Research. Within Bell Labs, this research organization is known simply as Research (with a capital R), or as Area 11, which is its organization number. Except in cases of ambiguity or when there is a need to emphasize, I will use the shorter Bell Labs or Research to mean Bell Labs Research.

NOT YET – THE JURY IS STILL OUT

Bell Labs has been called America's national treasure and the crown jewel of AT&T and Lucent. It has been a significant driving force in advancing science and technology in the twentieth century. Recently, Bell Labs has been receiving negative press because of Lucent's difficulties. For example, the August 9, 2001 issue of *Nature*[5] has an article about Bell Labs with the following title and highlight:

> ### Down and Out in Murray Hill
>
> The name Bell Labs is a byword for technological creativity. But its future is now clouded by the financial woes of its parent company.

Because of Lucent's difficulties, the future of Bell Labs may seem cloudy, and Bell Labs may be down, but it is not out. Skeptics believe that the best days of Bell Labs belong to the past, but others still have faith. For example, William Broad writing in the May 30, 2001 issue of *The New York Times* said the following:

> *Bell Labs and its parent company, Lucent Technologies, are still giants of innovation despite recent management fumbles and business failures, many industry experts say. ...*[6]

This faith in Bell Labs was based on its sustained record of decades of outstanding contributions to science. When I joined Bell Labs and for many years after that, it was a big deal to get a job there. Many of my friends and colleagues in universities were in awe of Bell Labs, a big and famous institution with a lot of history intertwined with the world of science. It would take me several years to get a good perspective of Bell Labs and its place in the scientific world.

The period of Bell Labs' contributions spans several more decades than my twenty-three years there. Bell Labs has a glorious history. Understanding it will give a better perspective of the challenges it faces now.

HISTORY

On January 1, 1925, AT&T announced the formation of a separate company, Bell Telephone Laboratories, as a joint venture of AT&T and its subsidiary Western Electric. AT&T president Walter Gifford appointed Frank Jewett as the first president of Bell Telephone Laboratories, in charge of about 4000 employees. Only a fraction of the employees were researchers, the others being engineers, systems engineers, and product development staff.

Bell Telephone Laboratories was informally known as Bell Labs, though the official acronym was BTL. The Telephone part in its name went away in 1984 when Bell Labs was officially renamed as AT&T Bell Labs.

Bell Labs started by taking over Western Electric's engineering department. Nine years later, in 1934, AT&T's development and research department was merged with Bell Labs to consolidate AT&T's research and development (R&D) activities in one organization.

Bell Labs began its operations at 463 West Street in New York City. The Murray Hill facility in New Jersey started operating in 1941. Eventually, Murray Hill became Bell Labs' primary research facility and its headquarters.

WHY CREATE BELL LABS?

In September 1924, AT&T president Walter Gifford[7] proposed the creation of a separate research entity for the AT&T companies, which were collectively known as the Bell System. He had three primary objectives for creating a separate research entity. First, Gifford believed that one central R&D facility would be more efficient and economical than having multiple R&D facilities. Second, the two R&D organizations in the Bell System, the engineering departments of Western Electric and AT&T were competing with each other for credit. If they were merged, this internal competition for credit would go away. Finally, a visible R&D organization would help to protect AT&T's monopoly status by demonstrating to its critics that AT&T was not only not stifling innovation but was actually encouraging innovation.

The idea of creating a separate R&D entity for the Bell System had started germinating many years earlier. John Carty, AT&T's chief engineer, strongly believed in the value of basic research. AT&T and Western Electric's R&D staff was spread across different geographical locations, in Boston, Chicago, and New York. In 1907, Carty started the process of consolidating R&D by moving R&D activities to New York. In 1911, Carty[8] was able to report the creation of a fundamental research group as part of Western Electric's engineering department. This group was set up so that it would have the best available talent and equipment to do the best research.

Carty's belief in the value of basic research convinced Theodore Vail,[9] two-time president of AT&T, to make basic research a major focus of AT&T strategy. In 1914, Vail wrote to the stockholders that 550 carefully chosen scientists and engineers were working in the Bell System headquarters in New York. Among them were ex-professors and holders of advanced degrees in science and engineering.[10] Vail appreciated the value of good research as a public relations tool to deflect criticism that AT&T's monopoly status was not good for the country. He set the precedent for AT&T executives, through the

AT&T monopoly era, to place more emphasis on science and technology than on profits in their public statements.[11]

Upon Vail's retirement in 1919, Harry Thayer succeeded Vail as the president of AT&T. Thayer had had a close business relationship with Vail, who was considered the Bell System visionary.[12] Thayer and Vail complemented each other, with Thayer being the implementer of Vail's visions, in this case making research an important part of AT&T strategy. Thayer, just before retiring in 1924, advised his heir apparent, Gifford, about the advantages of creating a separate AT&T research organization. Thus Gifford was acting on Thayer's advice when he proposed the creation of a separate research organization,[13] the next logical step.[14]

THE PRESIDENTS

Over its 75-year history, Bell Labs has had ten presidents:

President	Period
Frank B. Jewett	1925–1940
Oliver E. Buckley	1940–1951
Mervin J. Kelly	1951–1959
James B. Fisk	1959–1973
William O. Baker	1973–1979
Ian M. Ross	1979–1991
John S. Mayo	1991–1995
Daniel C. Stanzione	1995–1999
Arun Netravali	1999–2001
Bill O'Shea	2001–

The position of the Bell Labs president gradually became ceremonial after 1984 when Bell Labs became a division of AT&T and the development organizations were moved out of Bell Labs. The latter resulted in the Bell Labs president losing control of the huge development budget and also reduced his influence in setting the technology direction of AT&T. The responsibility for the divisions now within Bell Labs rests with their vice presidents. For example, the vice president of research is the head of Research.[15] Bell Labs presidents, if they so desire, have time to wear another hat in the corporate hierarchy.

The current (tenth) Bell Labs president, Bill O'Shea, is also Lucent's executive vice president of Strategy and Marketing, an extremely important position. O'Shea retained this position, which he held before becoming the Bell Labs president. Dan Stanzione, the eighth president of Bell Labs, was also Lucent's chief operating officer (and before that president of Lucent's Network Systems Group). Both Stanzione and O'Shea had their roots in the business units.

Arun Netravali, the ninth president, was a product of Bell Labs Research. Unlike O'Shea and Stanzione, Netravali is well known in the academic community and has received many awards. In addition, Netravali spent almost his entire Bell Labs career in Research. In contrast, O'Shea has had significant business experience in areas such as software development, manufacturing, marketing and sales, and acquisitions.

Unlike his predecessor, Stanzione, or his successor, O'Shea, being Bell Labs' president was Netravali's sole job. However, Netravali was not content with being a ceremonial figurehead. Consequently, he took an active role in managing research. His deputy, Bill Brinkman, the vice president of research, was a physicist, and Netravali was content to let him manage the physical sciences. However, it was a different story in the case of the non-physical sciences. Here Netravali set the agenda and made the key decisions. Netravali felt it necessary to do this since most of Brinkman's experience had been in the physical sciences. Netravali often worked directly with the managers and researchers in the non-physical sciences, helping to create projects and trying to establish collaboration with the business units.

Netravali was at the helm of Research from 1996 to 2001, first as vice president of research, and then as president of Bell Labs. The last two years of his tenure were during the difficult years for Lucent and, as a result, for Bell Labs. As the Research leader, Netravali was the first senior leader who proactively tried to change research direction and connect Bell Labs with the business units. Unfortunately, he had only limited success.

Circa 1984, Netravali was appointed director (now called research vice president) of the organization of which I was a member. We got to know each other soon because our kitchen was next to my office and Netravali would come there to get tea, which we both liked and drank often. He liked the location of the kitchen enough to have it remodeled into his office. Netravali was never my immediate manager. He was always a level or two higher than my managers. However, Netravali was more involved in my work than my managers. Consequently, I got to know Netravali well enough to even, occasionally, head to the nearest watering hole to chat and discuss work.

CONTRIBUTIONS TO SCIENCE AND TECHNOLOGY

Although most of us are not aware of it, an average home has at least 25 products that are based on Bell Labs technologies,[16] for example, telephones, TVs, faxes, computers, CD players, remote controls, and VCRs. Bell Labs' contributions to science and technology are simply enormous. Bell Labs has received nearly 30,000 patents, and has been filing about four patents each working day. Bell Labs' long list of inventions includes the transistor, the active communication satellite, the solar cell, the laser, the touch-tone phone, sound motion pictures, speech synthesis, the cellular telephone system, the UNIX® operating system, and the C and C++ programming languages.

Many Bell Labs inventions, for example, the transistor, the laser, and the communication satellite, have revolutionized society. Bell Labs researchers have also contributed to many other inventions that have led us to the Internet age. The inventions that influenced my professional career the most were the three software inventions, the UNIX operating system and the C and C++ programming languages, which represented Bell Labs computer science culture. Not only did many of us use them as tools for our research, but they also had a strong influence on the direction of software research at Bell Labs.

The research breakthroughs in Bell Labs took place because Bell Labs provided a great environment for research. Bell Labs gave its brilliant and hard working researchers access to leading technology and located together top-notch scientists such as physicists, computer scientists, mathematicians, material scientists, and system builders to intermingle and to cross-fertilize each other's ideas.[17]

For over three quarters of a century, in recognition of their contributions, Bell Labs scientists have received thousands of awards including Nobel Prizes, National Medals of Science, and National Medals of Technology. I, like most of my colleagues, was proud to be part of such an illustrious organization.

A GENEROUS AMERICAN INSTITUTION

Bell Labs (along with its parents, AT&T and Lucent) has been truly a generous institution in helping society. Bell Labs researchers and management have tried to help various groups that have been disadvantaged over the years. For example, Bell Labs has summer programs for minority college students, women graduate students, and high school students. Bell Labs also participates in Lucent's community oriented programs. The goal of these programs is to simply give the participants an opportunity to learn and make some money. Moreover, the old Bell System culture of helping society has led to many employees volunteering for community services.

Bell Labs has the tradition of offering a large number of summer jobs to students, often not expecting much in return for its investment. Every summer, many young persons can be seen in the Murray Hill hallways and in the cafeteria, reminding the researchers of university life.

Management has often gone out of its way to support universities. When a researcher was invited to give a talk at a university, Bell Labs would often pick up the tab. In addition, when faculty members wanted to visit Bell Labs, Bell Labs would often willingly agree to pay their expenses.

SIX NOBEL PRIZES & A TURING AWARD

Bell Labs researchers have won six Nobel Prizes in Physics[18] and one Turing Award,[19] which is a very remarkable accomplishment. I had never met a Nobel Prize winner until I came to Bell Labs. Nobel Prize winners are held in extremely high esteem, with almost god-like status, in many countries such as India, where I grew up. Consequently, I looked forward to meeting the in-house Nobel Prize winners when I joined Bell Labs. When I transferred to Research in 1979, a year after joining Bell Labs, one of my senior managers was Arno Penzias, winner of the 1978 Nobel Prize in Physics. Arno met all the characteristics that I had imagined in a Nobel Prize winner, for example, he talked with great confidence, was very eloquent, and even had a regal bearing. Over the years, I got to know Arno and he even agreed to write a foreword for my first book, which was on programming.

Although some of the Nobel prizes have been awarded for contributions that have had a great impact on society, it was the UNIX system, the subject of the 1983 Turing Award, that had a big impact on my research and writings.

WAVE NATURE OF MATTER

In 1923, the French scientist Louis De Broglie theorized that if light could sometimes be treated as a particle then perhaps any moving particle could sometimes be treated as a wave. In 1928, Bell Labs researcher Clinton J. Davisson experimentally confirmed De Broglie's theory by showing that electrons could be diffracted like light waves. Four months later, George P. Thomson (London University), using different equipment for his experiments, also independently confirmed De Broglie's theory.

Davisson and Thomson were awarded the 1937 Nobel Prize in Physics for demonstrating the wave nature of matter. Confirmation of the dual nature of matter has revolutionized theoretical physics and is an important foundation for much of current solid-state electronics.

TRANSISTOR

The transistor has revolutionized computing, communications, and all electronics. Every electronic device has transistors, and some tiny devices, such as microprocessors, have millions of them. Transistors can be used to open or close an electrical circuit, or to amplify electrical signals. Transistors are made from semiconductor material such as silicon or germanium and are microscopic in size. Currently, the size of the smallest commercially fabricated transistors is about 100+ nanometers and this is headed downwards. The smaller a transistor, the faster it can operate. The first transistor, fabricated in 1947, was about an inch long. (Claims made by Bell Labs scientists in 2001 that they had built a molecular size transistor, one nanometer in size, have turned out to be false.[20])

Bell Labs is where the electronic revolution started. In 1947, physicists John Bardeen, Walter H. Brattain, and William B. Shockley discovered the "transistor effect." That is, applying a small voltage to a semiconductor material can make it act like an electrical switch. Bardeen, Brattain, and Shockley also developed the first transistor. They were awarded the 1956 Nobel Prize in Physics for discovering the transistor effect. Although they had done their pioneering work at Bell Labs, by the time Bardeen, Brattain, and Shockley were awarded the Nobel Prize, Bardeen and Shockley had left Bell Labs. Bardeen was at the University of Illinois and Shockley was at Beckman Instruments.

The tiny transistor has changed the world in a big way! Reflecting on the invention of the transistor upon Shockley's passing away in 1989, *The New York Times*[21] commented:

The invention of the transistor became the basis for the electronic age. From it flowed virtually every one of today's devices installed in airliners and cars, calculators and computers, wristwatches and washing machines.

MAGNETIC AND DISORDERED SYSTEMS STRUCTURE

Unlike atoms in crystalline materials, atoms in "disordered" materials, such as glass, do not form lattices (regular geometric structures). The lack of lattices in disordered materials makes it hard to build a theoretical framework for understanding them.

Bell Labs researcher Philip Warren Anderson made significant contributions towards the understanding of disordered materials. Anderson's research led to an understanding of the electrical conductivity in glass. His research was a vital element in the understanding of magnetic phenomena in materials such as copper and silver, which in their pure form are not magnetic.

Anderson (Bell Labs), Sir Nevill Francis Mott (Cambridge University), and John Hasbrouck van Vleck (Harvard University) were awarded the 1977 Nobel Prize in Physics for their fundamental theoretical investigations of the electronic structure of magnetic and disordered systems. Their research was in the area of solid-state physics, which underlies the field of electronics.

BIG BANG

In 1948, George Gamow, the noted physicist, and colleagues theorized that the universe was created as a result of a cosmic explosion, called the "Big Bang," about 15 billion years ago. According to the Big Bang theory, very high temperatures of about 10 billion degrees Kelvin (°K) accompanied the creation of the universe. (The Kelvin scale starts at absolute zero while the Celsius scale starts at the freezing point of water, which is 273 degrees above absolute zero or 273°K. A change of one degree Celsius in temperature is equal to a change of one degree Kelvin.)

The high temperatures were required for building the chemical elements from elementary particles, which were assumed to be present from the very beginning, and they implied a very high amount of cosmic radiation. It was expected that this cosmic radiation would diminish with the expansion of the universe, which is another part of the Big Bang theory. It was theorized that if this cosmic radiation could be measured now, it would be equivalent to radiation emitted by an object at a temperature of about 3°K, that is, three degrees Kelvin, which means three degrees above absolute zero.

Scientists had long assumed that it would be impossible to detect and measure such a tiny amount of radiation because of the presence of cosmic "noise," that is, radiation from stars and other cosmic objects. However, in 1965, two Bell Labs radio astronomers, Arno A. Penzias and Robert W. Wilson, were able to detect and measure this radiation. The radiation was about 3°K, as predicted by the Big Bang theory, thus confirming the theory.

In detecting the radiation from the Big Bang, Penzias and Wilson used a very sensitive horn shaped radio antenna that was built in the early 1960s for communicating with AT&T's communication satellites Echo (passive type) and Telstar (active type). When this antenna was no longer needed for communicating with the satellites, Penzias and Wilson were able to use it for studying microwave background radiation, which led to the discovery of the radiation from the Big Bang.

Penzias and Wilson were awarded the 1978 Nobel Prize in Physics for their discovery of cosmic microwave background radiation. The Nobel Prize committee emphasized the importance of this discovery in its citation:

> *The discovery of Penzias and Wilson was a fundamental one: it has made it possible to obtain information about cosmic processes that took place a very long time ago, at the time of the creation of the universe.*[22]

Penzias and Wilson shared the 1978 Nobel Prize in Physics with Pyotr Leonidovich Kapitsa (USSR) who was honored for his basic inventions and discoveries in the area of low-temperature physics.

TRAPPING ATOMS WITH LASERS

Studying individual atoms and molecules in solids and liquids is difficult because they are packed too closely to each other. Atoms and molecules in gases are not packed so closely, but studying them is a problem because they move around with great speed. For example, atoms and molecules in the air move around at a speed of about 4000 km/hr.

When the temperature approaches absolute zero, the speed of the atoms and molecules diminishes significantly. For example, at a temperature of 10^{-6} degrees above absolute zero, the speed of free hydrogen atoms slows down to less than 1 km/hr. To avoid condensation and freezing, this cooling must be done with a few atoms in a vacuum. At this speed, individual atoms can be studied with extreme accuracy.

In 1985, Bell Labs researcher Steven Chu devised a technique that used lasers to lower the temperature to just above absolute zero. He did this by using three pairs of laser beams to slow (cool) down atoms in a vacuum. The lasers behaved like "optical molasses" in slowing the atoms.

Chu (Stanford University), Claude Cohen-Tannoudji (Collège de France, Paris), and William D. Phillips (National Institute of Standards and Technology, USA) were awarded the 1997 Nobel Prize in Physics for the development of methods to cool and trap atoms with lasers. Although Chu had done his Nobel Prize winning research at Bell Labs, Chu was no longer at Bell Labs. He had left in 1987 for Stanford University.

The Nobel Prize committee cited the work of these three Nobel Laureates as

*... helping us to study fundamental phenomena and measure impor-
tant physical quantities with unprecedented precision.*[23]

NEW PARTICLES WITH FRACTIONAL CHARGES

In 1982, Bell Labs researcher Horst L. Störmer and Daniel C.
Tsui of Princeton University showed that electrons in strong magnetic fields could associate to form new types of particles with frac-
tional electron charges. A few months later, Robert B. Laughlin, a
researcher at the Livermore Labs in California, was able to explain
their result.

Laughlin (Stanford University), Störmer (Columbia University
and Bell Labs), and Tsui (Princeton University) were awarded the
1998 Nobel Prize in Physics for their discovery of a new form of
quantum fluid with fractionally charged excitations.

The Nobel Prize committee described the significance of this discovery saying that the

> *Knowledge of electrons is a key to the events and the processes of
> our time. Soon the circuits used will be so small that quantum fluids
> with different types of quasi-particles may become forces to be reck-
> oned with in everyday computers.*

> *This discovery ... may well turn out to be essential for our future
> information society.*[24]

It is interesting to note that all the three Nobel Laureates had a
Bell Labs pedigree. Laughlin left Bell Labs for Livermore Labs in
1981 and Tsui left for Princeton in 1982. Störmer is now a Professor
of Physics at Columbia University, but he still holds a part-time appointment at Bell Labs. Störmer, talking about basic research at Bell
Labs, said

> *... it was my biggest desire in life to go to Bell Labs because Bell
> Labs is just the best place in the world to do the kind of research I
> wanted to do.*[25]

THE UNIX OPERATING SYSTEM

In 1965, Bell Labs and GE researchers joined the MIT Multics project to build the next generation operating system that could reliably serve thousands of users while running nonstop 24 × 7. However, a few years later, in 1969, Bell Labs withdrew from the Multics project because the project was not on track despite the investment of much time and money.

Bell Labs researchers then started exploring the building of another operating system as an alternative to the Multics operating system. Ken Thompson and later Dennis Ritchie wanted to build an operating system that would give them a pleasant environment for writing and using programs.[26]

After some discussions on what should be in an operating system, Thompson built an early version, soon to be named the UNIX operating system, on a PDP-7. However, a more powerful computer was needed on which to build a multi-user time-sharing version of UNIX as desired by the researchers. Bell Labs management rejected requests by Ritchie, Thompson, and others to purchase a computer in 1969 because it would cost over $100K.[27]

At a colleague's suggestion, the researchers convinced management to buy a PDP-11 for building a text processing system. Management found this proposition attractive because text processing in Bell Labs used significant resources. A large number of typists were employed for typing reports, technical papers, letters, and other kinds of documents. A text processing system could potentially save Bell Labs a large amount of money.

The PDP-11, which cost $65K, arrived in late 1970. The researchers' real motivation for purchasing the PDP-11 was to continue development of the UNIX system. According to Dennis Ritchie:

> *We knew there was a scam going on — we'd promised [to develop] a word processing system, not an operating system.*[28]

The UNIX system, along with a text processing application, was ready for real customers in 1971. It was successfully deployed as a text processing system for the Bell Labs patent department, the first UNIX customer.

Thompson was the key person behind the design and building of the UNIX system. Ritchie joined Thompson in building it. He also designed the C language, which was used for rewriting the UNIX system.

The UNIX system became very successful for a variety of reasons. For example, early versions of it ran on very popular computers, the PDP-11 and the VAX. The UNIX system was rewritten in the "high-level" language C, which made it relatively easy to implement on other computers. Universities embraced the UNIX system because AT&T offered them inexpensive academic licenses and gave them the source code. The University of California at Berkeley developed an enhanced version of the UNIX system, called Berkeley UNIX, which was widely used in universities.

The UNIX system was a big step forward in making computers user friendly. Operating systems of the time, such as the IBM's OS/360, were written in assembly language and required expert knowledge to use them effectively. In contrast, the UNIX system was relatively easy to use and modify because of innovations such as "pipes," the hierarchical file system, the user interface, and the fact that it was written in the C programming language.

Pipes provide UNIX users and programmers with an easy way of connecting programs to build complex applications. Using a pipe, the output of one program can be easily made to become the input of another program. For example, a program that produces as its output a list of misspelled words in a document can be used in conjunction with a sort program to list the misspelled words in alphabetical order without requiring a change to either program. Pipes are a key component of the UNIX philosophy of providing an environment to write small and specialized programs that can interoperate easily with other programs. To facilitate this process, UNIX programs are typically written to operate on text input streams and to produce as their result text output streams. Pipes can then be used to "glue" such programs together to perform tasks that are more complex.

Ritchie and Thompson were awarded the 1983 Turing Award for their development of generic operating systems theory and specifically for the implementation of the UNIX operating system. Thompson retired from Bell Labs in 2000 and is now a Fellow at Entrisphere, Inc.

Talking about the success of UNIX, David Tilbrook,[29] the founder of the first Canadian UNIX company, said that

> ...[the UNIX system] wasn't a great advance in computing; if anything, it was a great simplification, it put into the realm of the user things that were just inconceivable prior to that.[30]

The popularity of the UNIX system and its impact on computing has been phenomenal. The UNIX system put Bell Labs on the map in the computer world. A generation of computer scientists has grown up with the UNIX system. Hundreds of books have been written on or about the UNIX system,[31] including two by me that are about document formatting and typesetting on the UNIX system.[32] All sorts of machines run the UNIX system or its variants such as Solaris from Sun, AIX from IBM, or UNIX compatible systems such as Linux and FreeBSD.

By 1994 there were over 3 million computers running UNIX systems,[33] and this number is estimated to have risen to 4.5 million systems by 1998. This count does not include UNIX compatible systems such as Linux. By 1998, around 14 million computers were running Linux, and this number was headed towards 30 million by 2003.

BELL LABS FISSIONS

Between 1984 and when I left Bell Labs in 2001, many major changes took place that affected Bell Labs. For example, because of the 1984 AT&T divestiture, AT&T went from being a mega-giant monopoly to an "ordinary" telecommunications giant that had to compete for business. In doing so, AT&T spun off seven Baby Bells (or RBOCs, the Regional Bell Operating Companies), which were giant telephone companies in their own right. Bell Labs, which used to be a separate company, became a division of AT&T.

A dozen years later, AT&T split up into three companies. This three-way split was called the "trivestiture" as a play on the word divestiture. These breakups led to Bell Labs being handed to Lucent:

YEAR	BELL LABS OWNER	OFFICIAL NAME
1978 (I join)	AT&T (along with its subsidiary Western Electric) Logo color is blue.	Bell Telephone Laboratories, Inc. (separate company)
1984 (AT&T divestiture)	AT&T Logo[34] color is sky blue.	AT&T Bell Laboratories (becomes part of AT&T)
1996 (AT&T trivestiture)	Lucent Technologies Logo color is red.	Bell Labs (becomes part of Lucent)

Then in 2000, Lucent split up, spinning off two companies, Avaya and Agere, and selling many of its businesses so that it could focus on its core telecommunications equipment business.

With each split or sale of business, the business activities of Bell Labs' parent narrowed, which in turn reduced the research areas of potential interest to the parent. This impact was compounded by the fact that the telecommunications market first exploded and then shrank quickly, putting Lucent and the other telecommunications equipment makers in a very difficult situation. Lucent's difficulties led to the downsizing of Bell Labs and a shift of focus away from basic research towards working with the business units.

The AT&T and Lucent breakups caused Bell Labs to shrink because it had to spin off "baby" labs. Bell Labs progeny can be seen in the following "family" tree:

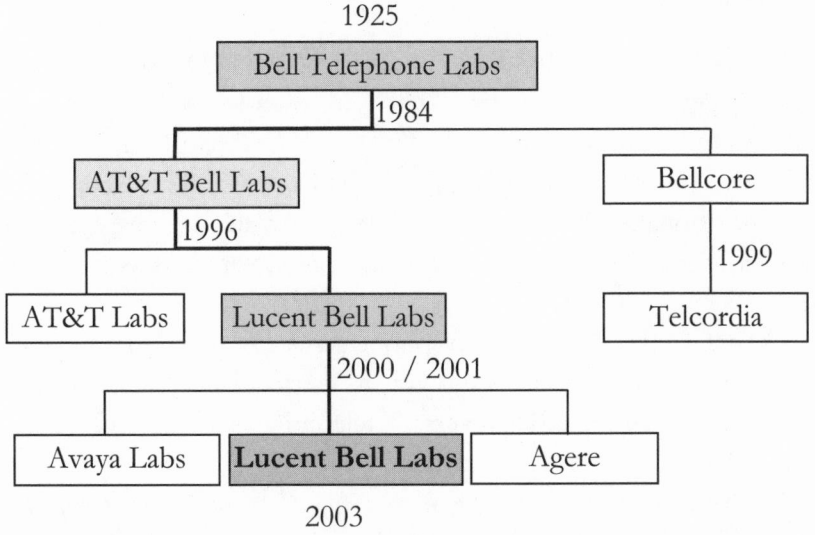

Agere does not have a separate research organization. Bellcore, the research lab for the RBOCs, was acquired by SAIC in 1997 and renamed Telcordia in 1999.

Every time Bell Labs was split, researchers usually had the choice of going to a baby lab or staying with Bell Labs. There were some exceptions. For example, organizations doing research specific to the business being spun off were assigned to go to the "baby" lab. In my case, in each of the three splits, I had the choice of going to a baby lab. There were pros and cons for staying at Bell Labs, but in the final analysis, staying at Bell Labs won in each case.

For example, in the case of the AT&T trivestiture, I felt that a systems and equipment company like Lucent was more likely to support research than a services company, which was what the new AT&T would become. Moreover, most of my researchers were involved in research projects that seemed to have a better with fit with the charter of Lucent. Some non-technical pros for staying at Bell Labs were its history of achievements, its global fame, and working at Murray Hill.

1982 AT&T CONSENT DECREE – DIVESTITURE

For decades, AT&T provided end-to-end telecommunication services and equipment to its customers. AT&T had been operating as a government regulated monopoly since 1913. The US government had bought AT&T president Theodore Vail's argument that the most efficient way of operating a telephone system was as a monopoly.

AT&T provided the best telephone service in the world and most of its customers were satisfied with its performance. Its rates were reasonable, it provided "universal" telephone service, and offered great customer service. By contrast, rates in Europe were much higher and, in some countries, it took a long time to get telephone service. For example, in India it took years, along with a substantial signup deposit, for a customer to get telephone service at all.

Nevertheless, some customers were unhappy because they felt that AT&T's monopoly status allowed it to charge long distance rates that were higher than warranted or that a competitive environment would allow. To some degree, these customers were right. AT&T was using a portion of the long distance rates to subsidize local telephone service and was, as a result, able to offer local telephone service to everyone at affordable prices.

In addition, potential competitors of AT&T were not happy at being left out of the big American telecommunications market. For many years they had complained to the government that the AT&T monopoly was stifling competition and innovation in telecommunications. Many potential competitors including MCI, the latter in the late 1960s, had been campaigning for loosening AT&T's lock on the telecommunications market.

In 1974, the US government filed an antitrust lawsuit against AT&T. After years of expensive litigation, in 1982, AT&T agreed to settle with the US government. One important motivating factor for AT&T to settle was that it wanted to get into the booming business of selling computers. AT&T was already making computers for its switches, but it was not allowed to sell them. As a regulated monopoly, AT&T could only be in the telecommunications business.

As part of the settlement, AT&T agreed to divest its subsidiaries that provided local and regional telephone service. AT&T would retain long distance service, telecommunications equipment manufacturing, and R&D, and it would be allowed to enter the computer business. The divestiture took place on January 1, 1984 with Ma Bell giving birth to seven regional operating companies, the Baby Bells.

Bell Labs stayed with AT&T, but the Baby Bells needed some R&D. To reduce R&D costs, the Baby Bells agreed to set up a shared research facility, called Bellcore, which they would jointly own and fund. AT&T agreed to seed Bellcore with a limited number of researchers by giving Bell Labs researchers the option of joining Bellcore.

Despite the change in its business climate, AT&T considered Bell Labs important for its future. Thus although AT&T shrunk by about two-thirds because of the divestiture, Bell Labs Research shrunk by only about 10%:

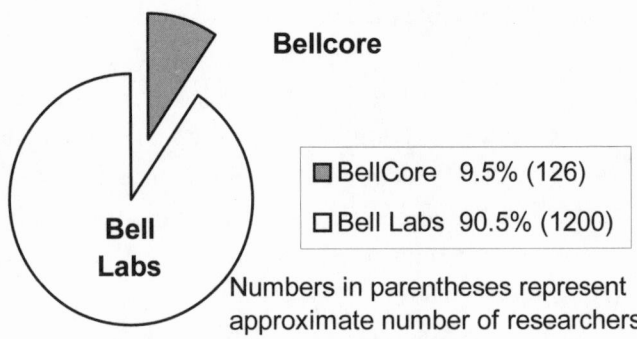

Bellcore

- ■ BellCore 9.5% (126)
- □ Bell Labs 90.5% (1200)

Numbers in parentheses represent approximate number of researchers

With AT&T becoming a competitive telecommunications company, Bell Labs Research now had to consider AT&T's business needs. In the monopoly days, funding Bell Labs Research was an expense item for AT&T that did not have an impact on its profits, which were guaranteed. But now AT&T would be funding Bell Labs Research from its profits, which would decrease correspondingly. Consequently, AT&T expected Bell Labs Research to deliver a return on its investment. In addition, Bell Labs Research (and the other organizations in Bell Labs) now needed to be accountable to AT&T management. For these and other reasons, the structure of Bell Labs was changed from that of an independent company to a division of AT&T.

Bell Labs Research management believed that AT&T's commitment to basic research was so strong that divestiture would not affect it much, if at all. After all, the Bell Labs Research budget would still be a very small part of AT&T's revenues. For example, talking about the post-divestiture Bell Labs, Kumar Patel, head of the Physics Division in Research said

> ... *those people who say Bell Labs has become more applied are better at measuring second and third derivatives than I am.*[35]

Around the same time, Patel's boss Arno Penzias, then vice president of research, said the following about Bell Labs' mission:

> *Only rarely will our work appear as a product before it is discussed in the scientific journals. One manner in which Bell Labs is likely to change is the speed and flexibility with which advances from the Research Department are explored for systems applications.* [36]

Patel and Penzias believed that Bell Labs would continue to operate as before, which was in line with the views expressed by AT&T management, for example, in the 1982 annual report addressing the consequences of divestiture:

> *... the management of AT&T has pledged to continue support of fundamental research at Bell Laboratories.* [37]

The Bell Labs Research focus on basic research and publishing continued for many years after AT&T yielded its monopoly status. Publishing was viewed as important for the professional recognition of researchers and for recruiting new researchers. As we will see, only in the 1990s did the focus of Bell Labs Research shift towards building systems and developing technologies and products for the business units.

MOVING THE "D" OUT OF BELL LABS R&D TO THE BUSINESS UNITS

To prepare for life after monopoly, AT&T was restructured so that it could operate efficiently as a competitive company. It was organized into business units, which were profit centers, and Bell Labs. The research part of Bell Labs constituted only a small part of its workforce, now about 1200 out of 18,000 employees. [38] The other organizations in Bell Labs were development organizations building systems for the business units. Since the business units were funding these organizations, they needed to be made accountable to the appropriate business units. Consequently, the development organizations in Bell Labs were moved to the business units. [39] Despite the move, these development organizations continued to use the prestigious Bell Labs name as an umbrella for their activities to ensure a continuity of culture and to help in recruiting.

Research (and some other central organizations) stayed in Bell Labs and would continue to be funded directly by AT&T as before.

AT&T Trivestiture

In the years that followed AT&T's entry into the competitive arena, it was becoming increasing clear to the AT&T leadership that there was not much synergy between its services and telecommunications equipment businesses. In fact, the service part of AT&T was hurting equipment sales. AT&T competitors in the long distance business, such as MCI and Sprint, were balking at buying telecommunications gear from AT&T. Buying from AT&T would simply be "aiding the enemy" by adding to its already huge coffers and giving it even more resources to compete against them.

The telecommunications equipment selling business was about to get more difficult for AT&T. Both AT&T and the Baby Bells, AT&T's traditional customers, were now positioning themselves to compete with each other. The Baby Bells were planning to enter the lucrative long distance market while AT&T was planning to offer local telephone service. As a result, the Baby Bells were increasingly buying equipment from AT&T's competitors such as Nortel and Alcatel.

One way to solve the customer conflict problems for AT&T's equipment business was to spin it off as a separate company. Then companies like MCI, Sprint, and the Baby Bells would have no hesitation in buying equipment from the new company since they would not be aiding their competitor AT&T.

On a different front, AT&T's foray into the computer business was not faring well. It had cost the company billions of dollars. The computer business was bleeding cash. To stop the flow of red ink, in September 1991, AT&T acquired NCR in a hostile takeover at a cost of $7.4 billion dollars.[40] AT&T then merged NCR with its faltering computer business in an effort to stop the bleeding. Despite the NCR acquisition and its subsequent merger, AT&T was still not faring well in the computer business.

Even though the above problems were public knowledge, AT&T CEO Robert Allen's announcement on September 20, 1995 that AT&T would be split came as a total surprise to most people. The goal was to unlock the value of AT&T's business by splitting it into three publicly traded companies. One company would be a systems and equipment company, which would eventually be named Lucent Technologies. Another would be a computer company, which would eventually retake its original name NCR and regain its status as an independent company. The third company would be a communications services company, which would keep the AT&T name and logo.

AT&T's voluntary break-up was the largest ever in the history of corporate USA. Lucent became a fully independent company on September 30, 1996,[41] and NCR regained its independent status on January 1, 1997.

The AT&T trivestiture was going to have a direct impact on Bell Labs, as it would now be split once again. AT&T and Lucent (but not NCR) would have research labs. Bell Labs would become part of Lucent, but about a quarter of Bell Labs' researchers would be assigned to the new AT&T Labs:

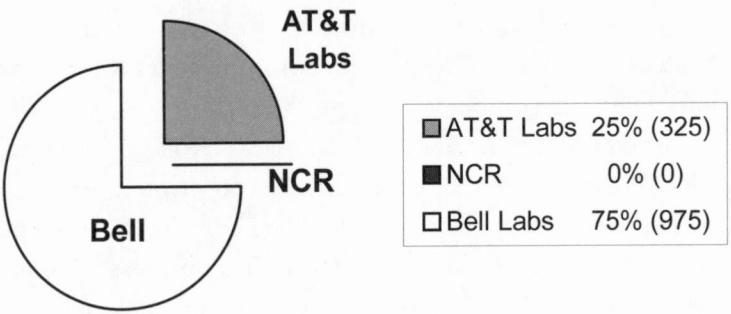

■AT&T Labs	25% (325)
■NCR	0% (0)
□Bell Labs	75% (975)

Splitting Bell Labs Again

Lucent would start its life as an independent company with about 140K employees, about 40% of the AT&T employees before its split. But Lucent's Bell Labs would keep about 75% of the researchers, the rationale being that research was more important for a systems and equipment company than for a services company.

NCR did not get any part of Bell Labs because it had been with AT&T for only a few years and there was no history of interaction between Bell Labs and NCR. Moreover, despite many attempts, mostly by Bell Labs, to collaborate with NCR, no substantial interaction had taken place.

LUCENT TRIVESTITURE

The rationale for AT&T spinning off Lucent seemed to be right on the money. AT&T competitors, who had not been buying equipment from AT&T, now started ordering equipment from Lucent and Lucent's sales soared in the early years. By January 2000, Lucent's stock had risen to a new high of around 80 dollars (not counting stock splits).

However, two months later, by March 2000, Lucent's stock had dropped into the 50s[42] because Lucent's revenues and earnings had fallen below market expectations. Ostensibly to address the sagging stock price, Rich McGinn, the now ousted CEO, said that he would spin off Lucent's non-core businesses in an effort to unlock the value of Lucent's assets. This would also allow Lucent to focus on its core telecommunication equipment business. The spinoff of Lucent's enterprise business was announced on March 1, 2000. The new company, which was named Avaya, started trading as an independent company on October 2, 2000. The spinoff of Lucent's microelectronics business was announced on July 20, 2000. This company, which was named Agere, started trading as an independent company on March 28, 2001.

Once again, it was time to split Bell Labs. Along with Avaya and Agere, went some of Bell Labs Research:

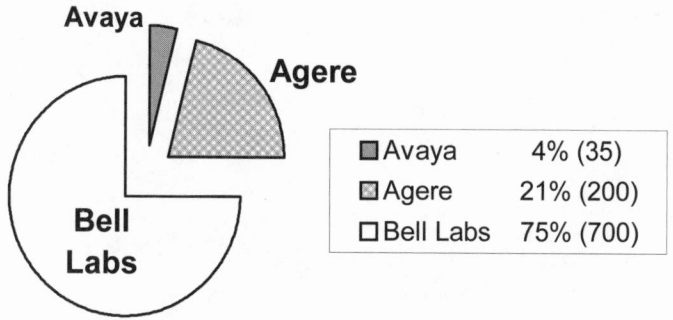

◼ Avaya	4%	(35)
▨ Agere	21%	(200)
☐ Bell Labs	75%	(700)

Splitting up Bell Labs Yet Again

DOWNSIZING

Towards the end of the twentieth century, the size of Bell Labs went down because its corporate parent shrunk by splitting, selling, and spinning off parts of its business. Unfortunately, the shrinking of Bell Labs did not end here. By early 2001, Lucent's financial situation was starting to become difficult. Budgets were being cut and the number of employees in Lucent had to be reduced. For that reason, the Bell Labs research facility in Silicon Valley was closed. By the summer of 2001, Lucent's financial situation had become more difficult. To cut costs further, more downsizing was needed. Bell Labs' remote locations in Greece, India, UK, and Illinois were closed. In addition, Bell Labs Research shrunk a bit more by "encouraging" voluntary retirements with incentives along with a few involuntary terminations (that came with a generous forced management package known as FMP). By late fall 2001, Bell Labs Research workforce was down to about 550:

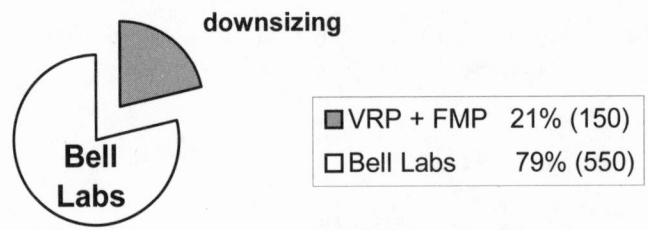

Downsizing Bell Labs

The downsizing of Bell Labs continued into 2002 with further cuts in the workforce bringing it closer to 500 employees. And there is talk of even more cuts.

LUCENT IS BORN!

The birth of Lucent was a very significant event for Bell Labs. There was euphoria in Bell Labs because it seemed that the top corporate brass had finally recognized the value of Bell Labs from a business perspective.

Lucent leadership did two things that cheered the researchers. First, they linked Lucent's name with Bell Labs by putting the endorsement line "Bell Labs Innovation" in the Lucent logo:

Lucent Technologies
Bell Labs Innovation

The logo unambiguously stated that Lucent was relying on Bell Labs technologies. Lucent products would be powered by Bell Labs' innovations, i.e., they would have "Bell Labs Inside."

The Lucent symbol was picked to be a bold, red, hand-drawn circle that was dubbed "the innovation ring:"

This again pointed to Lucent's association with the innovation factory called Bell Labs.

Of course, without a doubt, Lucent was capitalizing on the fact that the Bell Labs name was widely known and respected all over the world. Lucent executives realized that it would take time before the Lucent name would be able to match the Bell Labs name in brand value.

The second thing that Lucent leadership did that cheered the researchers was that it picked the main Bell Labs location at Murray Hill, NJ to be Lucent's headquarters. The senior leaders of the corporation owning Bell Labs would now be co-located with the Bell Labs researchers. They would now get an opportunity to see first hand the latest technologies and innovations being produced by the Bell Labs researchers.

In Bell Labs, both researchers and management hoped that Lucent's decision to make Murray Hill its headquarters would be more than a symbolic gesture and that it would help in building better bridges between Research and the business units. Co-location would make it easier for researchers to interact with Lucent's leaders. Researchers and research management would no longer have to drive to AT&T's huge and magnificent headquarters in Basking Ridge to meet AT&T leaders. Basking Ridge was about ten to fourteen miles from Murray Hill depending upon whether one took the back roads or the highway.

Putting the Lucent business leaders in Murray Hill did make a difference to Bell Labs. For the first couple of years, Henry Schacht (the first and third CEO), and Rich McGinn (the second CEO) often walked over to see Research projects and to talk to the researchers. Arun Netravali, then vice president of research, built a good relationship with them and that augured well for Bell Labs Research. I even had the opportunity to give a few presentations to Schacht and McGinn.

It was refreshing to see the top brass of Lucent in Murray Hill's corridors, dining room, and even the cafeteria. In my 18 years before the creation of Lucent, I had seen the CEO of AT&T exactly once at close quarters; that was when the then CEO, Robert Allen, came to Murray Hill to see a demo of Netravali's HDTV project.

AT&T was generous in funding Bell Labs, but until the late 1980s, they did not seem to care what Bell Labs did or did not do as long as they excelled in science. With Lucent, the situation seemed very different. Lucent, at least its top leadership, really wanted Bell Labs to help develop technologies for next generation Lucent products, for Bell Labs to be Lucent's innovation engine.

BELL LABS IN LUCENT

Bell Labs now consists of two main organizations, Research and Advanced Technologies, the latter being an internal contracting organization. In the past, Bell Labs also used to have many product development organizations, which were transferred to the business units as AT&T entered competition.

Lucent is organized as a bunch of business units and Bell Labs. By early 2002, after many acquisitions, spinoffs, sales of businesses, downsizing, and reorganizations, Lucent business organizations were reorganized into two business units, InterNetworking Systems (INS) and Mobility Solutions. INS supplied products and services to *wireline* service providers while Mobility Solutions supplied products and services to *wireless* service providers.

Trying to recover from its financial difficulties, Lucent has narrowed its focus to supplying products and services to about 30 global customers, all of which are large services providers.

Bell Labs researchers now work in areas such as computer science, communications software, physical sciences, data networking, and optical networking. Of course, the spectrum of each of the research areas is much narrower now than it was in the past, even a few years ago. With every breakup of its parent, the number of areas of interest to the corporation has shrunk and within some areas, the focus has become narrower.

Bell Labs' parent corporation has always centrally funded Bell Labs Research.[43] As a result, business units have never had direct control of Bell Labs' research direction, although they were able to influence it.

When Lucent was launched as an independent company, Lucent leadership assured Bell Labs management that the Research budget would be set at 1% of Lucent revenues. In the first few years, Bell Labs Research's budget increased in tandem with Lucent's increasing revenues. However, the dawn of the new millennium saw the Bell Labs Research budget decrease for two reasons. First, because of financial difficulties, Lucent reduced Research funding from 1% of revenues to 0.75%. Second, Lucent revenues dropped significantly due to the difficult market conditions (and due to spinoffs and sales of businesses), which automatically led to a reduction in the Research budget.

PILLAR OF
SCIENTIFIC & TECHNICAL INGENUITY

Bell Labs' record of scientific accomplishments is unquestionably stellar. William Broad wrote the following in *The New York Times*:

> *By nearly any standard, Bell Laboratories ... is a pillar of scientific and technical ingenuity.* [44]

Bell Labs is a household name in the scientific community and working at Bell Labs has been the dream of many a scientist. To scientists all over the world, Bell Labs has been the mecca for basic research[45] because of its brilliant scientists, scientists in multiple disciplines, accomplishments, and a wonderful research environment.

AT&T gave Bell Labs 10% of its R&D budget for basic research. Support for basic research was ingrained in the AT&T culture. For example, AT&T president F. R. Kappel[46] talked about basic research before the Economic Club of New York in January 1958. Kappel said that the Bell System had to keep acquiring new knowledge and this called for a "special brand of brains." Research should have an objective, but the "brains" must be given full freedom, which gives the researchers the opportunity to exercise their full creative powers. Researchers were thus free to select their own research topics without worrying about business relevance or management approval. They could even ignore management suggestions to stop working on a topic without the fear of serious negative repercussions, provided the research led to good results.

Bell Labs was like a university that had no students, a zero teaching load, no tenure problems, no running around for grants, and plenty of money for equipment and travel. A researcher could focus on building his or her professional credentials and reputation. Within a few years, with Bell Labs on his or her resume, the researcher would have a passport to a tenured position at one of the top universities or would be able to walk into a senior research position at one of the industrial research labs.

There was no real penalty for failure to produce research results in the short term, say a year or two. A researcher would not lose his or her job because of an unsuccessful or bad year from a research perspective. A researcher was therefore not penalized for working on a topic that led to a dead end. Lack of research success was often chalked off to bad luck, especially if the researcher had had successes in past years. Such researchers were even given a raise at the end of the year.[48] The rationale was to allow researchers to take risks and work on topics that could potentially lead to revolutionary, not just evolutionary, innovation. A researcher who made a habit of working on inappropriate topics or ones that led to blind alleys would be encouraged to leave but not fired.

Bell Labs thus offered its scientists an environment where they could think out of the box in the pursuit of innovation and invention. As a result, Bell Labs scientists came up with numerous inventions many of which, such as the transistor, the active communication satellite, and the laser, had a profound effect on society.

Researchers were also attracted to Bell Labs because its management welcomed and appreciated the external recognition earned by its researchers. In fact, until the 1990s, professional recognition was considered by (local) management to be more important than recognition within the company. This attitude was very unusual for a corporation because internal recognition came by doing things of value for the company's business and most companies wanted their employees to do this.

Researchers were rewarded generously for getting external recognition and fame. Most researchers worked long hours to earn such recognition and many researchers did become quite famous. Bell Labs researchers generated plenty of good publicity for AT&T (and later Lucent) with papers in prestigious conferences, patents, articles in newspapers and magazines, and the awards received by them. AT&T management was aware that such publicity was hard to buy with money.

According to Horst L. Störmer, 1998 Nobel Laureate in Physics:

> *[Bell Labs is] so good that everyone wants to go there. Over a very
> long stretch of time it was the best place in the world and it attracted
> – and attracts – the best people. The situation has somewhat changed
> nowadays ...* [48]

The situation has changed because the needs and the financial
condition of Lucent have changed. It used to be that AT&T was
happy to let the researchers reach out for the sky and make Bell Labs
famous. AT&T basked in reflected glory, which did not cost it anything in the monopoly days, since it was allowed to pass on the costs
to its customers. However, when AT&T shed its monopoly status,
the rules of the game changed, which meant that AT&T would have
to fund Bell Labs from its revenues, thus reducing its profits. Consequently, AT&T expected Bell Labs to help it compete by developing
technologies that would lead to new products and services.

Bell Labs started to slowly change its research direction beginning
in the early 1990s, shedding science for science's sake and focusing
on topics of interest to AT&T. Bell Labs also started trying to engage
the business units[49] with the goal of working with them.

In the first few years following its birth, Lucent performed ex-
traordinarily well financially. But the early years of the new millennium saw Lucent in financial difficulty, caused partly by the lack of
new products and the rest by market conditions. Lucent now needs
Bell Labs to help its business and do so urgently to give it an edge in
the difficult telecommunications market.

Within Bell Labs, internal recognition for helping Lucent business units has become more important than external recognition.
With the survival of Lucent at stake, contributions to the success of
the company have assumed paramount importance.

3 Life At Murray Hill

WALKING BACKWARDS

I HAD HEARD many a time, after joining Bell Labs in Piscataway, that researchers at Murray Hill were not only arrogant but also a bit strange. I was somewhat skeptical of such remarks since I felt that non-researchers did not quite appreciate the intensity, the focus, and the conviction with which successful researchers had to work. Moreover, I was also used to such remarks because they are occasionally also applied to university professors, and I had been a professor for a few years. A few months after I joined Bell Labs, the group I was in relocated from Piscataway to Murray Hill. Although I discounted what I had heard about the Murray Hill researchers, I was on the lookout for unusual behavior at Murray Hill when we moved there.

Nothing could surpass my amazement at what I saw at Murray Hill during one of my first few days there. As I was walking back to my office, I saw a person walking backwards, walking close to the wall in the middle of a long narrow corridor. The backward walker was a bit crouched and he was looking over his shoulder to avoid obstructions and colliding with other people.

I was extremely surprised. I could not believe what I was seeing. Here is a person, probably a brilliant researcher, walking backwards, in the heart of the world famous Bell Labs that had produced many thousands of inventions including several that had changed the world radically. Bell Labs Research was apparently stranger than I had thought it might be and the folks at Piscataway were probably right after all.

Why would a researcher be walking backwards? Perhaps he had made a wager with a colleague that he could walk the whole length of the long corridor backwards, without colliding with anything or anyone, and without feeling embarrassed. I let the incident pass without trying to find out more about the backward walker. However, a few days later, I saw the same person, in the same corridor, walking backwards again. Now I felt that there was more to backward walking and that it was probably not related to a wager. I decided to find out more about the backward walker by asking a few senior researchers, who had been around at Murray Hill for some years.

As it turns out, I was completely wrong in my speculation about the motivation of the person for walking backwards. He was a well-respected researcher who, because of serious knee and balance problems caused by severe injuries from an automobile accident, found it less painful and much easier to walk backwards.

I JOIN RESEARCH

Most people around the world think that all of Bell Labs is involved in research. Certainly, that is what I used to think when I joined Bell Labs in 1978. Actually, as I learned later, research was a very small part of Bell Labs. Bell Labs Research had about 1300 employees. The total number of employees at Bell Labs circa 1978 was around 25,000. Most of Bell Labs consisted of organizations that did development projects for the AT&T business organizations.

The distinction between development organizations and Research within Bell Labs was not explained to me, and, since I did not know any better, I had not asked. I was hired to do research in programming methodology. As a result, I thought I was joining Bell Labs Research. But, in reality, I started my years at Bell Labs by joining a development organization whose charter was to develop a version of UNIX, called the PWB/UNIX, which provided tools for developing large software systems.

The development organization I joined was in Piscataway, NJ. There were thousands of Bell Labs employees in Piscataway. However, all the Bell Labs organizations in Piscataway were development organizations. There was no research organization in Piscataway.

Bell Labs employees in Piscataway talked about the researchers as the learned folks up at Murray Hill working on esoteric topics, most of which had nothing to do with AT&T's business. Rarely did the Murray Hill researchers wander over to Piscataway and rarely did the developers in Piscataway head over to Murray Hill. There was little or no interaction between the developers at Piscataway and the researchers at Murray Hill with one exception (to my knowledge). The PWB/UNIX group, for obvious reasons, did interact with the UNIX creators and other researchers involved with UNIX at Murray Hill.

I soon realized that my research interests to explore issues in programming language design did not match the needs of the PWB/UNIX development organization, which was focused on building a product. Except for a few of us, everyone else was working on building something for PWB/UNIX and no one besides my manager cared about what I did or did not do. I had joined Bell Labs to do research not software development. Therefore, I decided to try to transfer to Bell Labs Research in Murray Hill.

Transferring to Research from a Bell Labs development organization was considered to be next to impossible in the 1970s. Once a person joined a Bell Labs development organization, the person was "tainted for life" in the eyes of the researchers. Such a person would find it very hard to transfer to Research, even if he or she had excellent qualifications.

Despite the skepticism of my co-workers, I applied to transfer to Research. In 1979, after many months of waiting, I was able to transfer to Research in Murray Hill[50] primarily because of Dhiraj Sharma, a researcher I had met at the seminars on new technologies that I had been organizing at Piscataway.

When I joined Research, I met my manager John Limb for a few minutes and that was about it. He welcomed me and I was on my own after that. My understanding with him was that I would spend 50% of my time investigating the very new area of office automation, whatever that meant. For the remaining 50%, I could work in any other area of my own choosing. This 50% commitment to work on office automation was the price I had to pay for being able to transfer to Research from a development organization.

Limb's attitude towards his research staff was in line with the Bell Labs philosophy, which was to hire the best people, get out of their way, and let them do research, publish papers, and become famous. In the words of Dennis Ritchie, the author of the C programming language and the winner of the Turing award for his work on UNIX, the Bell Labs philosophy was and is to

> ... *hire good people who generate their own ideas and carry them out* ... *The details and flavor of the grot vary between academia and [Bell Labs] but the ultimate expectation is surprisingly similar. People are given a lot of freedom to blaze their own path.* [51]

Until the late 1980s, Bell Labs cared only about academic research excellence. Bell Labs respected and valued its researchers for their innovations and their contributions to science. Nobody in Bell Labs cared whether or not a research topic was motivated by a business need or was relevant to AT&T. William O. Baker, the fifth president of Bell Labs, believed that in their quest for innovation, researchers would be motivated by factors other than the business needs of AT&T. He realized and appreciated the fact that the fundamental drive among top-notch researchers would be their own view, not AT&T's view, about the importance of their work. A researcher's loyalty to his or her scientific community could be as strong or even stronger than loyalty to AT&T.[52]

Research at Bell Labs was about advancing science, not about developing technologies for AT&T products. According to Baker, Bell Labs scientists were trying to find paths in nature, in science, in understanding, which would involve the deepest intellectual insight and skill, and which would simultaneously be of great benefit to mankind.[53]

BELL LABS WAS FAMILY

Joining Bell Labs was like becoming a member of a family. Besides work related interactions, there was much social interaction between the employees, both researchers and the non-researchers, at Murray Hill. Many employees would go for lunch in groups, go for after lunch walks around the Murray Hill grounds, or play a few games of bocce during lunch. In my first few years, some of us, along with our significant others, would also get together for lunch on Saturdays at the nearby Chinese restaurant.

Bell Labs was instrumental in creating a family-like culture by encouraging and supporting extra-work activities for its employees. Bell Labs had clubs for activities such as yoga, skiing, hiking, cinema, public speaking, tennis, softball, karate, and tai chi. In addition to allowing the clubs to use its facilities, Bell Labs also funded some of their activities. Usually, the club activities happened during lunch, in the evenings after work, and on weekends. However, all through the workday, some employees could be seen playing chess on the second floor just above the cafeteria entrance. Out of sight, squirreled away near the attic in one of the buildings, there was a table tennis table where researchers could be found playing, usually during lunch or after work.

The Murray Hill Cinema Club showed movie classics every other week, on Friday nights in the huge Murray Hill auditorium. In the 1980s, about a dozen of us would regularly go to the Bell Labs movies. Before the movie, we would gather at the house of John Linderman (now with AT&T Labs) for dinner. He lived within walking distance of Murray Hill and was very generous in hosting us for dinner every other week.

Motivated researchers worked long hours pursuing that elusive link in a theorem or an algorithm, or finding the elusive bug so that they could make their systems work correctly. Researchers could be found working at Murray Hill all the time. On weekends and holidays, a lonely researcher could always find someone at Murray Hill to talk to, or have lunch or dinner with. There were researchers at Murray Hill late at night, early morning, before dawn, on weekends, on Christmas Day, on New Year's day, and so on. Some researchers would come late in the afternoon and work late into the night. Other would come early in the morning and stay until late in the evening. Some would work from home part of the week. Some researchers worked "normal" hours while others worked unconventional hours. Management did not care when and where a researcher worked as long as research publications and peer evaluations showed that the researcher was doing good work.

THE RESEARCH & INVENTION FACTORY

Having spent three years teaching at a university and about a year or less in a Bell Labs development organization, I found Bell Labs Research to be a very stimulating environment in comparison. There were several reasons for this. Bell Labs had researchers exploring a wide variety of scientific areas, which created a fertile environment for the cross-fertilization of ideas. Bell Labs had many world-renowned experts and chatting with them required only walking over to their offices. Most researchers were willing to discuss research problems and were often eager to collaborate. They were ready to comment on papers written by their colleagues. The researchers were usually very candid in giving feedback to their colleagues, which would help improve the quality of their papers.

Finally, AT&T was very big and involved in so many products and services that there were always real-world problems to drive research projects. For example, the research in billing and customer care that was started by researchers in my department in the early 1990s was motivated by the needs of the company. These projects led to patents, research publications, and, most importantly, the building of innovative systems that were potentially valuable for AT&T.

In many ways, Bell Labs Research was better than a university, especially for those interested in pursuing a research career and who did not care about teaching. Bell Labs Research gave its researchers all the freedom they wanted and the funds to do world-class research. Bell Labs Research itself was organized into departments, much like university departments. The departments were discipline oriented and not oriented in accordance with the company business or its products and services. For example, Bell Labs had departments that did research on disciplines such as distributed systems, programming languages, and computer systems. However, until the 1990s, typically research departments did not focus on business topics such as PBXs, billing, or telephone services.

Department managers, called department heads (now called directors) played a role similar to that played by the chairpersons of departments at universities. Just as chairpersons wield little power over the professors, the department heads wielded little power over the researchers. The researchers were like professors, but without the chores that would distract them from focusing on investigating research problems, producing results, and publishing them.

The one very big and visible difference between Bell Labs and a university is the lack of students. Many researchers would welcome graduate students at Bell Labs. Graduate students are a key component of research projects at universities. They can be extremely motivated and are "free" resources that a professor can leverage to advance his or her research.

Another difference was that at a university, a professor could be a big fish in a small tank, while at Bell Labs a typical researcher would be a small fish in a big tank. Of course, the famous researchers, and there were many of them at Bell Labs, were big fish in a big tank.

Bell Labs had plenty of resources and, unlike universities, no paucity of funds for research activities and equipment. For example, in the late 1970s, most professors were still punching cards to write programs. At Bell Labs, the world of interactive computing had arrived, and each researcher had his or her own computer terminal. In the universities, professors had to be careful about making long distance calls. For obvious reasons, at Bell Labs, researchers could make essentially unlimited long distance business calls. A researcher could talk with colleagues all around the world without having to worry about the minutes ticking by.

Murray Hill has an excellent library. Unlike many university libraries, it was very well staffed with reference librarians and research specialists. The library has most of the latest technical books and journals. Before the Internet, if a researcher needed a book from the library and did not feel like walking, all he or she needed to do was to telephone the library and request the book. The book would be sent to the researcher by intra-company mail and it would be at the researcher's desk within a few hours (in the 1970s and early 1980s, there was internal mail delivery four times a day). If the book was not locally available, the librarian would get it from one of the other Bell Labs library locations and the book would be in the hands of the researcher within a day or two. If a researcher wanted to buy a book that the library did not have or just because the researcher preferred to have his or her own personal copy, the researcher could order the book through Bell Labs. Researchers could buy all the books they wanted for their research at Bell Labs' expense. In fact, many did take advantage of this perquisite and built extensive collections of books.

Bell Labs researchers did not have to rely on students to build systems. Students are not only transient, but most of them also do not have the skill set to build robust commercial quality systems (of course, there are some notable exceptions since some tremendous systems have come out of universities). Bell Labs researchers could request their managers to provide them with a programmer. Programmers for working on research projects were a limited resource. Consequently, to get a programmer, a researcher had to be on good terms with his or her manager or, alternatively, the researcher had to convince the manager about the importance of the research project.

Collaborating with colleagues was easier at Bell Labs than at universities, since unlike university professors, researchers generally came to work for the whole day. A researcher could count on his or her colleagues being available most of the time. In universities, colleagues can disappear in the summer months and every now and then, they go away for yearlong sabbaticals. Such absences tend to disrupt collaborations.

Bell Labs paid researchers extremely good salaries, which could be much higher than academic salaries. For example, in 1978, my starting Bell Labs salary was 60% more than my university salary (nine month salary prorated to an annual salary).

UNIVERSITY-LIKE INSTITUTION

For decades, AT&T was content to let Bell Labs Research operate like a university because it did not cost AT&T anything. The money AT&T spent to run Bell Labs came from the telephone users in the form of slightly higher rates. As a regulated monopoly, AT&T was guaranteed a minimum profit. The regulatory agencies allowed AT&T to set rates that would take into account the expense of running Bell Labs. Thus, AT&T charged its customers higher rates with the full blessing of the US government, which was happy with the scientific advances emanating from Bell Labs and the perception that Bell Labs was a national treasure.

Plenty of money, a collection of extremely bright researchers, freedom to select research topics, and the freedom to do research without worrying about business needs set Bell Labs on the road to glory in the world of science and technology. The glory allowed Bell Labs to morph from an industrial research lab to a university-like research lab during AT&T's monopoly days. This change caused Bell Labs to become, according to Robert Buderi,

> ... *increasingly disconnected from the rest of the company, partly as a result of the sweeping success of its scientific investigations that transformed a corporate laboratory into a university-like institution.*[54]

By the time I joined Bell Labs Research in 1979, there was very little interaction between Bell Labs Research and AT&T's business divisions because AT&T did not need Bell Labs to protect and advance its business interests. The US government had taken care of that by making AT&T a government-protected monopoly. Consequently, AT&T was content with the publicity that Bell Labs that was generating for it frequently and in large quantity. Bell Labs researchers were also happy at not having to worry about AT&T's business needs, which allowed them to focus on science and technology.

AT&T's shedding of its monopoly status started to swing the pendulum in the direction of industrial research, away from university-style research. Although AT&T and Bell Labs management did not publicly acknowledge it, the divestiture was the signal for Bell Labs to morph back into an industrial lab. This was going to happen, despite management's initial belief that not much would change at Bell Labs. AT&T would now be funding Bell Labs with "real" money. Sooner or later, especially when the business climate got tough, AT&T would expect Bell Labs to deliver a return for its investment.

THE RESEARCHERS

Bell Labs researchers set very high standards for themselves and for their colleagues. They placed a lot of emphasis on originality, invention, quality, and long-term research. As far as the researchers were concerned, their world consisted of Bell Labs Research and their peers in universities and other industrial research labs. Others in AT&T and Bell Labs, those outside of Bell Labs Research, did not really matter. The researchers essentially had no contact with anyone in AT&T sales or marketing and little contact with the development organizations. In fact, many researchers looked down upon sales and marketing activities since they believed that good technology would sell itself.

Many researchers were very candid and blunt in technical discussions. They had no qualms about telling a colleague that his or her work was not up to snuff, for example, the approach was wrong, the work was not innovative, the premise was speculative, and so on. Such researcher openness and frankness was acceptable to the researchers and managers within Research, but others, such as the business unit colleagues, interpreted it as researcher arrogance.

The researchers, even the world famous ones, always had time to discuss research problems with their colleagues and were often willing to collaborate. All my significant research projects involved voluntary participation of my colleagues without which the projects would not have gone very far. For example, my key collaborator in Concurrent C/C++,[55] the parallel version of the programming languages C and C++ that we designed, was Bill Roome. He became interested in working with me on Concurrent C/C++ after we discussed some of my parallel programming ideas. Similarly, my key collaborator in starting the Ode object database project[56] was Rakesh Agrawal, the resident database guru. He became interested in working with me after I shared with him my idea of storing C++ objects in a database.

Selecting a research topic at Bell Labs was similar to selecting a PhD thesis topic at a university. A good topic had to be challenging, be of current interest to researchers both inside and outside Bell Labs, and one that would lead to publications and possibly to an interesting prototype (in case of researchers who were system builders). Bell Labs hired promising researchers because Bell Labs considered their areas of expertise to be scientifically important. Researchers were generally not hired because of some specific business need. They were also not hired to work on a specific project or on a specific topic. Consequently, researchers could work on any topic that grabbed their fancy, mildly subject to the charter of their organizations. However, Bell Labs was usually flexible enough to even allow a researcher to work on a topic outside the scope of his or her organization's charter.

The research environment at Bell Labs was intentionally quite unstructured. Once hired, researchers were left to fend for themselves. To start with, the researchers had to find research topics on their own. They would not get much guidance from management in selecting research topics, and each researcher would have to swim or sink with his or her choice of a topic. Most researchers liked the freedom to select a topic, but some new researchers found this freedom combined with the lack of early management feedback very disconcerting. As a result, some of them would flounder and could become problems for management.

Bell Labs management typically did not tell the researchers what they should work on nor did they officially bless the topics selected by them. Consequently, researchers could not blame research failure on management and they had to accept responsibility in such cases. Many managers considered researchers who could not find their own research topics not to be Bell Labs material!

Most new hires at Bell Labs were freshly minted PhDs, who would typically join in the summer or early fall following graduation. Employee performance reviews, an annual exercise, were held at the end of the year. Thus, the first performance review of the new hires would happen about three months after they joined. Since this was not enough time to evaluate them, new hires, especially fresh PhDs, would get a "pass" to the next performance review. This gave the new hires about fifteen months to establish their research program before their performance was evaluated. The second performance review was critical for the new hires since it would set the tone for their career at Bell Labs.

Publishing papers in conferences and journals is an important activity for researchers because it represents validation of their work by their peers and it gets them professional recognition. Consequently, Bell Labs encouraged its researchers to publish papers in top-notch conferences and journals. In fact, in the 1980s, research publications were a very important component of a researcher's performance evaluation. From the Bell Labs perspective, publications gave it visibility and it helped in recruiting other researchers. Bell Labs had a very liberal publication policy. Even though the papers had to be cleared by the publications review board before they were submitted for external publication, approval was granted routinely. The publications review board would solicit the approval of the business units before approving papers for publication. The goal of the approval process was to ensure that no company proprietary information or information that would give competitors an advantage was published. During my 23 years at Bell Labs, I heard of only a handful of cases in which the business units objected to the publication of a paper because it contained proprietary information. In most cases, business unit objections were resolved after some negotiations, which sometimes required modifying the papers in question.

The research mantra at Bell Labs used to be long-term research. Almost everything else was too pedestrian for the researchers. They did not like to be involved with any project that smacked of short-term research or was being driven by the business units. Some researchers used the long-term research mantra as a way of avoiding accountability and not having to show results. Long-term research, they would argue, meant that it would be years before they could show results.

However, all this started changing in the 1990s, especially after the birth of Lucent. Researchers, many of whom had traditionally never worried about the business value of their projects, were now being urged to work on projects with high business value. They were told that it was time to move away from projects with low or no business value. On the topic of business value, I had an interesting chat with a manager, whom I will call Joe. Like many others, Joe had a passion for long-term research, research that was unencumbered by business pressures. My chat with Joe went something like this:

> **Narain:** I would like to try a business exercise with you. Your current budget is about $6 million. What value are you producing for Lucent in return for the $6 million that Lucent is investing in you and your team?
>
> **Joe:** It's hard to quantify the value we are producing for Lucent. As you know, the value of long-term research is hard to quantify in the short term. Your emphasis is on working with the business units and short-term projects. It's easier to measure value in such cases. Many of my team members and I are working on long-term research projects. It will take about three years before these projects produce results. Only then will we be able to determine the value of my projects.
>
> **Narain:** Fine, I understand your difficulty. Humor me by doing the follow-

ing. Let us take the researchers who
have been working on long-term research projects for the last three
years. What value has their research
produced for Lucent?

Joe: (*silence*)

Narain: Incidentally, because we are
looking at a three-year time span,
we have to consider the total amount
that was invested in these projects
over three years.

Joe: This is a new way of looking at
long-term research for me. I need
time to reflect on how to answer
your query.

In the last few years, Bell Labs researchers have become very aware of the need to work on topics of direct interest to the Lucent business units, topics that will give Lucent a competitive edge.

OUTLETS FOR RESEARCH TECHNOLOGY

AT&T, and later Lucent, business units were the channels for taking Bell Labs technologies to the marketplace in the form of products and services. However, most researchers often selected research topics with little or no understanding about the needs of the business units. The researchers had little or no interaction with the business units until the 1990s. It was hard for the researchers to get the attention of the business units since, for cultural and historical reasons, they were predisposed to not working with Bell Labs. Because of the "disconnect" between Bell Labs and the business units, research projects were rarely part of the product roadmaps (plans) of the business units or their strategy. Consequently, most research projects had little chance of making it to the marketplace.

In the 1990s, a few senior managers such as Netravali attempted to rectify this situation. Several projects were started at Netravali's behest, based on his understanding of the company's business needs. Netravali also tried to get the business units to work with the researchers, but here his success was very limited because he had no direct control over them.

Netravali, to his credit, was always trying to move Bell Labs in the direction of the company business. He often encouraged (coerced) researchers to work on specific projects. Despite their lack of business understanding, researchers did not like management to tell them what they should work on. Researchers considered Netravali's projects to be development not research. As a result, Netravali earned the ire of some researchers who felt he was bent on destroying long-term research.

In the case of software, the problem of taking the systems developed by the researchers to the marketplace was even more difficult than doing the same for projects in other areas such as networking and speech. In the 1980s, Bell Labs started expanding its software research program by hiring many researchers in software disciplines such as operating systems, programming languages, databases, and compilers. Here, Bell Labs was ahead of AT&T, which was not yet in the software business although there was serious talk about it entering the software business. AT&T business units were channels for telecommunications products and services, but not for software products. The business units preferred to sell the familiar telecommunications products and services in which each new deal would bring them hundreds of millions of dollars. Software was too small a market to catch the attention of the existing business units.

In 1991, AT&T, eager to expand its footprint in the computer business and at the same time stop the red ink flowing from its existing computer business, acquired NCR. The NCR acquisition made Bell Labs researchers think that AT&T was serious about the software business and that they would now have a software channel. Unfortunately, NCR did not turn out to be a software channel. It did not have a research facility of its own and its staff was not used to working with very independent researchers like those at Bell Labs. In any case, NCR's focus was on system integration and not on making and selling new software products.

The AT&T trivestiture in 1996 separated Bell Labs (and Lucent) from NCR and any remaining hopes about NCR being a software channel. Researchers next hoped that Kenan Systems, acquired by Lucent in 1999, would be their software channel. Kenan Systems was in the billing and customer care systems business and their chief, Kenan Sahin, was very interested in marketing software. However, the researchers' renewed hopes for a software channel was short lived because Lucent decided to exit the billing and customer care systems business two years later in 2001.

RESEARCHERS AND CUSTOMERS

Until recently, business units rarely involved Bell Labs researchers with their customers. Most Bell Labs researchers and managers did not have the opportunity to meet customers. However, the situation is changing now as Research and the business units move towards working together more closely than they used to.

Unlike most researchers and managers, I did have the opportunity of dealing with many customers. They were customers of Maps On Us (www.mapsonus.com), the Web business that my colleague Bill Roome and I created. I had the opportunity to make sales calls, win some deals, understand what it means to provide customer support, and so forth. Running Maps On Us gave me business experience, which I feel is very important for Bell Labs senior management to have. Technically speaking, since Maps On Us was a Lucent business, its customers were Lucent customers. However, Maps On Us was not affiliated with any Lucent business unit and its customers were not customers of Lucent's core telecommunications business.

Many researchers have little understanding of how difficult sales can be. Research is difficult, but this difficulty pales in comparison to the business difficulties like the ones faced by Lucent in dealing with the challenges of a very weak market for telecommunications equipment.

Researchers often champion new technologies without understanding the business costs and benefits of deploying these technologies. Contrary to what some researchers might think, customers are not looking for new technology just because it is new. Instead, they are looking for solutions that will enhance their businesses. Since most Bell Labs researchers in the past did not get an opportunity to meet and work with customers, they were not able to translate customer needs and requirements into selecting research projects and planning the direction of their research.

THE MANAGERS

Bell Labs has had a long tradition of egalitarianism. Rank is not an issue when it comes to research and technical opinions. Researchers express their opinions freely in meetings even though the opinions may differ from those of the managers present. This would, in all likelihood, not lead to any negative repercussions. Similar scenarios were less common in the business units because they were rank conscious. In addition, many researchers have free access to senior managers including the vice president of research and, in some cases, even the president of Bell Labs.

The wonderful thing about being a research manager was that there was tremendous freedom in doing the job. Essentially, each research manager was able to run an independent organization the way he or she thought appropriate. Talking about his job as the vice president of research, Arno Penzias said

> *Nobody tells me how to do my job, but there is tremendous pressure to do a good job.* [57]

The following figure shows how Bell Labs Research is organized and what its management hierarchy looked like in the fall of 2002:

President, Bell Labs

President, Bell Labs Research & Advanced Technologies
(Vice President of Research)
Organization Unit: Area

Research Senior Vice President
(Executive Director)
Organization Unit: Division

Research Vice President
(Director)
Organization Unit: Center or Lab

Director
(Department Head)
Organization Unit: Department

Researchers and other staff members

The titles in parentheses are the old titles, but they are still occasionally used informally. The new titles, such as director and research vice president, conform more to industry practice than their old counterparts. The old title, department head, which was used until a few years ago for first-line research managers, illustrates the extent of academic influence at Bell Labs. Departments have, on the average, about ten to fifteen researchers. A center typically has about five or six departments. A division has about four centers.

The top two Bell Labs leaders circa 2002 were Bill O'Shea, president of Bell Labs, and Jeff Jaffe, president of Bell Labs Research & Advanced Technologies.[58] They took over the reins of Bell Labs in the fall of 2001 from Arun Netravali and Bill Brinkman. Netravali had been the vice president of research from October 1995 to October 1999 and the president of Bell Labs from October 1999 to October 2001. Netravali is now Lucent's chief scientist. Bill Brinkman had been the vice president of research from January 2000 until his retirement in September 2001.

In early 2001, as research vice president, I reported directly to Brinkman since the research senior vice president position above me had been vacant. In the summer of 2001, Jaffe, then vice president of Advanced Technologies, and Brinkman's peer, was given the additional position of research senior vice president. In this capacity, Jaffe reported to Brinkman and I reported to Jaffe.

THERAPEUTIC VALUE OF RESEARCH

Until the 1990s, the primary role of management was to provide administrative support to the researchers so that they could focus on research and excel, which they did in plenty. First-line managers had little control over what the researchers did or did not do, especially in the case of well-established researchers.

Directors, the first-line managers, used to be like super researchers, i.e., researchers who also controlled the resources allocated for their departments. In fact, until very recently, most Bell Labs managers, particularly directors and research vice presidents, spent a substantial portion of their time on research besides managing their organizations. Some senior leaders also spent a substantial portion of their time doing research. For example, Netravali, even when he was the vice president of research and the president of Bell Labs, would spend a substantial portion of his time doing research in topics such as image processing and communications. According to Netravali,

> *I'm in management, but I still have my personal research programs. Right now, I'm spending 25 to 30 hours a week doing my own research.* [59]

Some researchers wanted to become managers, even though they disliked managing, because of the prestige, money, and the power associated with being a manager. They would choose the management path because Bell Labs did not have a technical ladder. Many managers viewed management tasks as chores to be avoided if possible, or to be done quickly, so that they could get back to doing research. They were most comfortable when doing research or writing papers. Research seemed to have a therapeutic value for the Bell Labs managers, similar to the relaxing effects of meditation, in getting rid of the stresses produced by management activities.

LIVE AND LET LIVE

Managers had a "live and let live" attitude with respect to the researchers, which was fine in the monopoly days when it was okay to let the researchers do whatever they wanted. The transition from managers as super researchers to managers helping develop new technologies and products for the company business started in the late 1980s, led by fresh thinking on the part of Bell Labs' leaders. For example, in 1989, Arno Penzias, vice president of research, realized that Bell Labs' research model was outdated. Instead of enjoying music at a concert, Penzias was fretting that there was something very wrong with Bell Labs Research. It was during the musical performance that Penzias came to the realization that Bell Labs needed to change with the times. Instead of writing and publishing papers or building the world's smallest lasers, researchers should be helping AT&T with its business objectives.[60]

Many Bell Labs managers (and researchers) were not eager to work with the business units because they found research to be far more intellectually challenging and satisfying than working with the business units. The latter required attending many meetings, which were often long and which usually made slow progress. Moreover, there was not much incentive for working with the business units nor was there a strong penalty for not doing so. Therefore, the managers applied little or no pressure on the researchers to change their research direction.

Netravali was the first senior leader to apply significant pressure on the managers and researchers to work with the business units. He pushed the managers to direct the researchers to work on projects with potential business value such as those in wireless and data networking and in areas new to Lucent such as network call centers and storage area networks. As was to be expected, Netravali encountered resistance from some managers and many of the researchers. They wanted complete freedom to work on whatever they wanted without being constrained by business needs.

Unfortunately, even if the researchers worked on business related topics, there was little possibility that the business units would actually use the results of their research. For example, in the late 1980s and early 1990s, I had led two successful research projects, building the Concurrent C/C++ parallel programming language[61] and the Ode object database.[62] However, they were not marketed as products because AT&T, despite its good intentions, had still not established channels to market software products. Nevertheless, I was convinced that if our research represented a good business proposition, then, with a little bit of luck, we would be able to partner with one of the business units who would find our research to be good from a business perspective and market it. Consequently, as a manager I encouraged and challenged my researchers to work on topics of value to the business units. For my efforts, I was criticized by some researchers and colleagues as being too product oriented and not being supportive of long-term research.

With the new millennium finding Lucent facing new challenges, the need for Bell Labs to work closely with the business units to develop new technologies and products became critical. Bell Labs leadership started looking for ways of making Bell Labs more relevant to the Lucent business. For example, in 2001, Bell Labs managers, led by the current research chief Jaffe, started the process of developing a formal and detailed research strategy plan to address Lucent's business needs. This strategy plan aims to focus resources on technologies that are important for the Lucent business units. Successful execution of the strategy will require the managers to guide and lead their researchers to work on topics important for the Lucent business. Bell Labs managers will also have to invest time and effort in developing better working relationships with the business units. Most importantly, the strategy plan has to have the involvement and buy-in of the business units.

NOT EVERYONE WANTS TO BE A MANAGER

Many Bell Labs researchers have over the years turned down promotion opportunities to become managers simply because they were very happy doing what they did best, that is, research. They just did not want to do the tasks associated with management. For example, Ken Thompson, the creator of UNIX, preferred to focus on technical work and never became a manager. Even though he was not a manager, Thompson wielded tremendous influence in the company, was respected by his colleagues, and was probably paid an excellent salary commensurate with his contributions. Similarly, Bill Roome, my collaborator in building the Concurrent C/C++ programming language and the Maps On Us Web service (www.mapsonus.com), decided that he enjoyed technical work and that he would not be a good manager. Consequently, he turned down several promotions. Nevertheless, colleagues and managers, including Netravali, respected Roome for his tremendous system building expertise.

However, many researchers who did not want to become managers nevertheless wanted public recognition for their contributions and the perquisites associated with becoming managers. A technical ladder would have addressed these issues, but Bell Labs did not have one. Senior Bell Labs management was not keen on the idea of a technical ladder, perhaps because it would require company wide approval and implementation.

Of course, becoming a manager meant better compensation. Superstar researchers, like Thompson and Ritchie, were in most likelihood paid more than many managers since there was no upper bound on researcher salaries. But, on the average, managers were paid more than researchers.

Unlike Bell Labs, the business units were very rank conscious and rank could be an impediment in dealing with them. In the business units, not being promoted within a reasonable period could mean that the employee was at best mediocre. The business unit managers preferred to deal with their counterparts in research or their seniors. Consequently, researchers who turned down promotions were generally at a disadvantage in dealing with managers in the business units.

RECRUITING THE BEST

The selection process for hiring new candidates at Bell Labs was very rigorous. Most researchers were hired from prestigious universities, came with very strong recommendations, and had published several papers even by the time they interviewed at Bell Labs. Recruiting was done primarily by word of mouth and by personal contacts.

Consider, for example, the research department named

Database Systems Research Department[63]

of which I was the director for some years starting in the mid 1990s when it was created. The department was very well known in the database community since it had many well-known researchers working on very interesting research projects. Researchers were publishing papers and building systems of potential value to AT&T (e.g., the Ode object database, the Columbus on-board navigation system, the Sunrise real-time billing system, and the Stair9 Web-based customer care system). Because of our high visibility in the database community, many top-notch PhD students would of their own volition send us job applications.

The hiring process at Bell Labs was similar to that followed at universities. Candidates, after much due diligence, were invited for a one- or two-day interview. The interview process would start with a seminar where the candidate would describe his or her research. Then the candidate would meet individually with interested researchers and managers. The candidate would also informally meet groups of researchers and managers over lunch and dinner. Each center[64] interested in hiring the candidate could make an employment offer. In case more than one center was interested in hiring the candidate, the "winning" center was determined by matching the interested centers with the candidate's preferences. The "offer" process was coordinated by the human relations (HR) organization.

PHDS VS. NON-PHDS

Most Bell Labs researchers have a PhD, but there were some notable exceptions. For example, Ken Thompson, the creator of the UNIX system did not have a PhD.[65] Most technical employees in Bell Labs Research without a PhD were programmers, not researchers. Hiring a candidate without a PhD was difficult and it required special approval of senior management. This preference for PhDs was another similarity that Bell Labs Research shared with the universities.

Many of these programmers, the so-called non-researchers, were extremely good and critical for many research projects. Their contributions were important in and essential to building many of the research prototypes and most of the few commercial quality software systems developed in Research. The programmers complemented the skills of the researchers.

Many researchers at Bell Labs do not have the skills or the experience required to build commercial quality systems. The researchers were hired, not for their programming skills, but for their ability to come up with new ideas and publish research papers. When urged to build real systems, some of the researchers were very vocal in saying that they did not come to Bell Labs to become software developers. If they were forced to become software developers, they would rather work for a software company where they would be appreciated for developing software or work for a startup where they would have a chance of making big money (this was in the late 1990s in the heyday of the dot-coms).

With Bell Labs moving quickly in the late 1990s to help develop technologies and products for the business units, it was necessary for Bell Labs to shift gears from publishing papers to system building. This made it somewhat easier to hire programmers and even researchers without a PhD.

STARTING SALARY ANOMALY

Salaries for new hires are computed by HR using a formula that takes into account items such as university degrees, area of expertise, salaries paid by competitors, and the number of years worked. The formula approach occasionally led to anomalies. Consider, for example, two equally qualified candidates graduating with a PhD in the same research area from the same university, but they differ in one way. One candidate has worked for a few years before starting his PhD, experience that is not directly relevant from a research perspective. However, the HR formula would give a higher salary to the candidate with the work experience.

I had one case similar to the above scenario. Increasing the inexperienced candidate's salary by a small amount would not eliminate the discrepancy. Increasing it significantly was not appropriate. We could decrease the salary proposed by HR for the experienced candidate, but we usually did not decrease the salary numbers given to us by HR.

Some months after the two researchers were hired, because of an administrative error, the researcher with the lower salary learned of the salary difference. Now this researcher was actually from a more prestigious university and was doing better work than the researcher with the higher salary. Understandably, he was livid. It took me many weeks to calm him, explain how the salary for new hires was determined, and assure him that I would try to remedy the discrepancy in the coming performance review. In the review, I explained the situation and we were able to fix the salary discrepancy by giving the aggrieved researcher a good raise.

GUIDING & CAJOLING THE RESEARCHERS

Starting in the early 1990s, some managers started encouraging researchers to work on topics with business value. In line with Bell Labs tradition, managers were at first tentative in urging these researchers to change research direction and work on topics relevant to the AT&T business. Consequently, most researchers continued to operate as before, business as usual, working on topics that had not much to do with AT&T's business.

For Bell Labs to deliver value to AT&T, both managers and researchers needed to change their *modus operandi*. Managers needed to, if necessary, direct the researchers to work on topics of value to AT&T. Researchers needed to be receptive to management suggestions about research direction.

The contradictory public and private stances of senior Bell Labs management on what was expected of the researchers confused both the researchers and their managers. Publicly, into the early 1990s, the position of the senior Bell Labs managers continued to be that the researchers should do world-class research, publish papers, etc. The focus was still not on working on topics with business value or working with the business units. Privately, however, senior management had very different expectations. They expected the researchers to work on topics relevant to the business units, preferably jointly with the business units. Researchers working with the business units were rewarded with better performance evaluations, higher raises, and stock options. These two stances were indicative of senior management's concern that a public shift away from basic research to helping the business units would have a serious negative impact on researcher morale and on Bell Labs' global reputation.

Because of these contradictory messages, researchers were torn between trying to establish themselves in the professional world and responding to subtle messages to work with the business units. Researchers had for decades shunned the latter as not being research. Unfortunately, many managers were not in a position to help the researchers much. They did not know much about the company's products or the business units, and they had few connections with the business units. On top of all this, many managers had only a limited understanding about the research areas of some of their employees. Until the 1980s, according to Penzias, almost everyone worked on a different topic, so there was no coherence.[66] A manager's expertise was often very different from that of the researcher.

Consider, for example, my manager in the late 1980s. He was an electrical engineer, not a computer scientist. He knew a lot about data networking and was well known for his contributions, but his expertise was not in software or databases. I was starting to work in the area of object databases, a new research topic that would complement Bell Labs' creation of C++. My manager tried to help by advising me to not work on object databases because from his perspective the topic was not new and therefore not worth pursuing. In any case, I chose to ignore his advice and we went on to build a pioneering object database, the Ode object database. I was able to ignore my manager's advice because Bell Labs gave researchers the freedom to ignore their managers, just in case they were wrong. However, in such cases, it was important for the researchers to be right and successful.

Breaking with tradition, in the 1990s, some managers including myself became very proactive in guiding researchers to work on topics that were both innovative and relevant from a business perspective. When necessary, I would cajole or even pressure the researchers to change their research direction. For example, I cajoled a researcher to work with our business colleagues on improving the order fulfillment process for the 5E switch (a big Lucent telephone switch), which used to take several months. The goal was to reduce this time to a couple of weeks. Despite the researcher's initial reluctance to work on this project, which would take him away from focusing on publishing, the researcher became quite involved in the 5E project and turned out to be a key contributor. One of his contributions was to build a workflow system that automated part of the order fulfillment process. The project was very successful and the researcher's contributions were duly recognized.

The standard initial reaction of most of my researchers when I encouraged them to change research direction was to inform me that I was micromanaging and not supporting long-term research. My response was that the issue was not long-term research, but working on the right topic. If they were not on the right track, it was my job to get them on track. In the final analysis, most of them realized that I was looking after their best interests and let me know their appreciation.

MAIN-MEMORY DATABASE: PAPERS OR A SYSTEM?

In the early 1990s, we hired two very bright computer scientists from prestigious universities. Both were excellent system builders, but at that time they wanted to focus on theoretical database research because that would lead to more publications and do so fast. Building real systems requires a lot of time and effort and involves a lot of pedestrian work. Moreover, when building systems, there is not much time left for writing papers and also there are generally fewer things to write about. Consequently, researchers who want to focus on publications often work on topics that do not require real system building.

The two new researchers were the leaders of a project to build a main-memory database that was just starting. An important factor in database performance is the speed of the disk, which is where the data is stored. Accessing data that is kept in main memory is much faster than accessing data kept on disk. Consequently, database performance can be significantly improved by keeping all the data in main memory and using the disk only for storing a permanent copy of the data. The idea of a main-memory database was not new, but the decline in memory prices had made main-memory databases economically viable. Building a main-memory database required developing new database algorithms that would be optimized for manipulating data in main memory.

As an ultra-fast database, a main-memory database had commercial potential, since there were many real-time critical applications that could benefit from such a database. Moreover, there was not much competition in this arena. Such a database could also be used in AT&T products to replace proprietary main-memory storage systems (not full-fledged databases). With a main-memory database, AT&T products could standardize on a single standard ultra fast database thus reducing costs.

Besides these two researchers, other researchers were also involved in the main-memory database project. The researchers wanted me to hire full-time programmers (one first, more later) to build the main-memory database. None of the researchers, including the two leaders, were keen on spending time actually building the main-memory database. Their preference was to do the research needed for developing the new algorithms, leaving the system building to programmers hired for the purpose. This would give the researchers time to write papers and work on other projects.

Seeing a lack of strong commitment on the part of the main-memory database team, I refused to hire programmers for the project unless and until the researchers made a commitment to work full time on the main-memory database project. Besides developing new database algorithms, I wanted them to be personally involved in building the main-memory database. Without total researcher commitment, the main-memory database would at best lead to some research publications, but it would not become a real system with real users and customers. In other words, unless a solid main-memory database was built, it would not have much business value for AT&T.

The researchers were vocally unhappy with me for wanting them to focus on building the main-memory database. Once again, the researchers accused me of not supporting long-term research and micromanaging. Nevertheless, I was adamant. No commitment, no programmers. Fortunately, the two leaders soon came around to my point of view and started focusing on building the main-memory database. Soon I hired a programmer to work with them. The main-memory database project started humming along at a fast pace.

A few months later, the two researchers, along with the main-memory database project, were transferred to another department. Their dedication was instrumental in moving the main-memory database towards becoming a product. The development responsibility of the main-memory database was eventually taken over by the Software Products Group business unit. The main-memory database was subsequently used in some Lucent products. However, the challenges faced by Lucent in the new century derailed the main-memory database project since Lucent decided to focus on its core telecommunications business. As a result, the main-memory database no longer matched Lucent's business interests. This led Lucent to sell the main-memory database to a software company.

The researchers, in addition to building the main-memory database, also published about ten research papers on the main-memory database technology developed by them. A real system and numerous research papers! Not many research projects in Bell Labs have produced such results. I was very gratified when one of the two research leaders came back some years later and told me that he appreciated my efforts in pushing him to focus on the main-memory database.

MCI's Billing Plan Drives Research

Inderpal Singh Mumick is one the brightest computer scientists that I have ever worked with. As a researcher, I would have found it tough to compete with him. Fortunately, I was his manager and the issue of competing with him did not arise. In his early years at Bell Labs, Mumick was publishing many papers in very prestigious conferences. In the mid 1990s, Mumick and a couple of other researchers started a real-time billing project named Sunrise. They were motivated to look at billing in response to MCI's "Friends & Family" program, which was taking away customers in large numbers from AT&T. This MCI program offered callers discounts for calls made between members of the same "Friends-and-Family" group. Thus, there was an incentive for group members to entice their relatives and friends to join their group.

A real-time billing system, such as Sunrise, that could tell users their current, up to the minute, billing charges and allow them to select or change billing plans in real-time would be a big coup for AT&T. Sunrise would be able to handle the "call detail" records for the about 200 million calls per day that were being made by the tens of millions of AT&T customers. The number of call detail records is much more than the number of calls since multiple call detail records are generated for each call.

Like the other Sunrise researchers, Mumick was working part time on Sunrise since he was simultaneously working on another project and was busy writing research papers. I urged him to work full time on Sunrise since real-time billing could potentially be very important for AT&T. I knew that by telling Mumick to focus on billing, I was taking a big risk, since this could make him unhappy and angry with me. However, success with Sunrise would be a big win for Mumick and the others and for AT&T. As I expected, Mumick was reluctant to work full time on Sunrise because of his other commitments. However, I was persistent. Fortunately, after some discussions, I was able to entice Mumick to focus on Sunrise. Eventually, the other researchers also started focusing on Sunrise.

With Sunrise, Mumick and the other Sunrise researchers laid the foundation for a real-time billing system for AT&T. Sunrise, circa 1995, became the responsibility of another manager because of reorganization. By this time, Sunrise was on its way to becoming a product, but the 1996 AT&T trivestiture came in the way. Sunrise's two partners, the Business Communications Services and the Consumer Communications Services business units, stayed with AT&T. However, Bell Labs went to Lucent. Sunrise stayed with Bell Labs, but two key researchers, including Mumick, went to AT&T Labs. Losing the two major partners, who were also potential future customers, and the two researchers was a setback for Sunrise. A couple of years later, Sunrise found a champion in Kenan Systems, the billing and customer care company that had been acquired by Lucent in 1999. Sunrise, renamed QTM (Quick To Market), now advanced towards being used in billing products. However, in 2002, Lucent exited its billing systems business by selling it to CSG Systems.

Mumick and the other Sunrise researchers had been very creative in coming up with interesting ideas about billing, which also led to several research publications. Mumick's departure was a loss for Bell Labs. Mumick eventually went on to form Savera, a Web-based billing company, and then Kirusa, a developer of multi-modal infrastructure for wireless applications.

I GET FLAK FROM THE RESEARCHERS

In the fall of 1997, we hired a very promising young researcher who joined Bell Labs at the end of the year. To address Lucent's business needs better with his research, the researcher decided to move away from his PhD research area and explore new topics, especially those that would be relevant to Lucent. Several months went by with the researcher working very hard, but without success in identifying a "good" research topic. The annual performance review, which was coming up in two or three months, posed a problem with respect to the young researcher. By then, the researcher would have been at Bell Labs for almost a year and there was a good chance that he would have no significant results to show for his efforts. If the researcher were tagged as a non-performer, then that label would be hard to erase. Despite his lack of progress, I felt that the researcher had too much promise to have a negative label attached to him.

Around this time, we were starting a directory (database) integration project to help the Lucent BCS division (now Avaya) improve its product offerings. If this researcher joined the team, he would be working on a project of real value for Lucent and would, at the same time, be exposed to new research opportunities. The researcher was not enthusiastic about working on the directory integration project because he felt that it would not be research. He did not agree with my view that real projects can lead to interesting research problems.

I was persistent in urging the researcher to work on the directory integration project. As a result, several of the researcher's colleagues came to argue with me on his behalf and I ended up taking a lot of flak from them. They berated me for asking a young researcher to work on a business related topic. One of them argued that the young researcher ought to be given at least two years, even three years, to establish his research program before Bell Labs evaluated his performance. Perhaps such a scenario would have been fine in the monopoly days, but in the current environment, it was important to deliver value much quicker.

I stood firm and eventually the researcher agreed to work on directory integration. This project was very successful and went on to become part of an Avaya product. While working on directory integration, the researcher got a chance to understand the Lucent business, work together with several colleagues, and publish a couple of research papers.

75% THINK THEY ARE IN THE TOP 25%

Researcher performance was formally evaluated once a year. A committee consisting of all the directors in a center and their vice presidents would collectively evaluate the performance of all the employees in their center.[67] Each director would represent his or her researchers in the review process, which could last up to two long days, to review about seventy five or so employees.

The review process was kicked off with each employee being asked to submit a completed accomplishment form listing his or her accomplishments for the year. The items listed would include research contributions, publications, systems built, awards, etc. Researchers were expected to use one page to describe their accomplishments. However, some researchers would use multiple pages to describe their accomplishments with the hope that more text was better than less.

Many researchers viewed filling out the accomplishment form as a chore. Researchers doing excellent work could get away without having to write much in their accomplishment forms. For example, in 1999, the Lucent Softswitch[68] project was generating a lot of attention and visibility. The two persons instrumental in creating the Lucent Softswitch were Murali Aravamudan and Shamim Naqvi, both directors. According to Naqvi, Aravamudan and he submitted identical accomplishment forms, which was very unusual in itself, with just a short single line containing words to the effect

I built Softswitch!

Despite their brief and identical accomplishment forms, Arava-mudan and Naqvi reportedly did very well in their performance review. Obviously, as should be the case, what mattered really was accomplishments and not simply the contents or the length of the accomplishment forms.

After evaluating the performance of the researchers, managers ranked or grouped them, for determining raises, by comparing their performance. Comparing the researchers was far more difficult than comparing apples and oranges because the comparison involved many different kinds of contributions in different and unrelated areas. Moreover, many of the evaluations and comparisons involved "give and take" on the part of the managers.

In the performance review, managers would detail the contributions of their researchers and try to ensure that they were rated fairly. Occasionally, a manager's claims about a researcher's contributions would become the subject of a disagreement. One year, for example, one manager claimed that one of his researchers was doing fundamental and pioneering work in software engineering and that his research had saved a business unit (about) $1.5 million. The manager argued for an excellent evaluation for the researcher. Many of us were skeptical of the manager's claims since the researcher was considered technically weak. To substantiate his claim, the manager whipped out a note from a business unit manager stating that the researcher's contributions had saved his business unit $1.5 million. We knew that such notes were not hard to get. It did not cost the business unit manager anything to write such a letter. Our boss, the presiding research vice president, was not sure about how to handle the researcher's evaluation because of the note.

At this point, I made the following suggestion. Since the researcher had saved the business unit $1.5 million, we should ask the business units to funnel some of the savings back to the researcher. Specifically, we should ask the business unit to hire a contract programmer for the researcher's project. The programmer would cost the business unit about $150K for one year, which was only 10% of what the researcher's work had saved them. With the programmer, the researcher might be able to save the business unit even more money next year.

As I expected, the researcher's manager was not happy with my suggestion. Nevertheless, I insisted that the researcher be given a good performance evaluation only if the business unit was willing to backup their savings claim with real money. That would give us confidence in believing that the $1.5 million savings actually happened. The manager was supposed to contact the business unit and get their response. This did not happen and we did not hear any more about the $1.5 million savings claim.

The review process was quite subjective out of necessity because there were very few objective measures available to us. It was not possible to translate each research contribution into dollars and cents. In the late 1990s, to make the review process objective as far as possible, some of us started using an employee evaluation matrix similar to the one shown below. Using such a matrix was not part of any standard Bell Labs employee evaluation process, since each group had its own way of evaluating researchers.

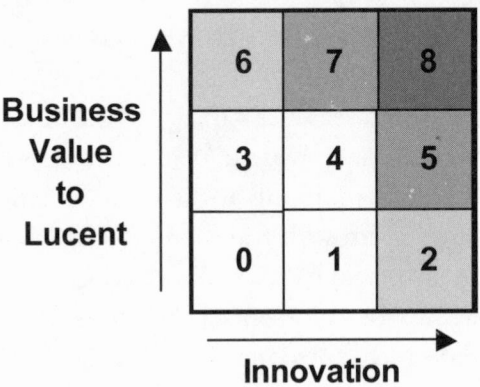

Unlike in the past, business value of the researcher's work would now be considered explicitly. The matrix had business value to Lucent and researcher innovation as its axes. I often used a matrix like this to explain to my researchers what was expected of them in terms of performance.

After a manager detailed a researcher's contributions and gave an assessment of the researcher's strengths and weaknesses, we would discuss the researcher's performance and then, by consensus, place the researcher in the one of the above squares. Employees in the shaded squares were the ones valuable to Bell Labs, with the darker shades representing more value. Obviously, our hope was to have more employees in the shaded squares and our goal was to help move them towards the darker shades, i.e., squares 5, 7, and 8, with 8 being the best scenario.

The squares were interpreted in a manner similar to that shown below:

SQUARE	COMMENTS
8	Extremely valuable employee. Innovative and working on topics relevant to the company business. Only a few employees fit this category.
5, 7	Valuable employee.
6	Does valuable stuff for the company, but not innovative.
2	World-class researcher not doing anything for the company. Goal is to help the researcher align with company interests. If the researcher has been in this category for many years, and is unwilling to do work that is relevant to the company, then the researcher should perhaps be told to find a job elsewhere.
4	Somewhat innovative, somewhat valuable to the company.
1	Typically, a fresh PhD who has not established himself or herself. The researcher should move to square 2 or better yet to square 5 in the next year. If the researcher is still in this category after a couple of years, then he or she should be asked to leave.
3	Somewhat valuable to the company, replaceable by a more qualified person.
0	Should be asked to leave.

Very few researchers would fall into square 8, and the rest of the top researchers would fall into the squares 5 and 7. Top programmers (as opposed to the researchers) would fall into squares 6 or 7. Unfortunately, as a legacy of past hiring criteria, which focused on publications, many senior researchers fell into square 2 and some into 1.

Up to the mid 1990s, most research projects were small, typically involving one or two researchers. The research culture was to keep projects small. Management did not encourage large projects involving many researchers because the managers received more credit for more projects in their organizations. According to Penzias, vice president of research, having a wide range of activities allowed the managers to hedge their bets. They would jump on the bandwagon of the project(s) that succeeded.[69] Another reason why managers did not encourage large projects was that they, being hands off managers, found it hard to determine the contributions of the team members and the significance of their contributions. As a result, in large projects, the better-known researchers would get most of the credit since that was the easiest way out for the managers.

Giving feedback to researchers about their performance was an important part of the performance evaluation. Telling a researcher that he or she had fared poorly in the performance review was difficult for some managers, especially since the researcher's area of expertise could be very different from that of the manager. Many managers found it easier to tell the researchers that they were doing just fine, even though they might have received mediocre evaluations. Many years ago when I was a researcher, we used to jokingly say that management likes to make 75% of the researchers believe that they are in the top 25%!

From a researcher's perspective, a better indicator of performance was the salary raise. Even though the salary raises did not have a big spread until the early 1990s, the raises could confirm or contradict the "feel good" written or verbal evaluations given by the managers. A raise close to or below the inflation rate was a good sign that the researcher was not doing well. Unfortunately, a researcher could not use the raise information to compare his or her performance with that of the others since the raises were private information.

Until recently, explicitly firing a researcher for poor performance was a rare phenomenon at Bell Labs. Usually, a researcher, upon learning that he or she was not performing well, would make plans to leave Bell Labs. In such a case, Bell Labs was generous in giving the researcher ample time, even a year or more if necessary, to find another job. Occasionally, a researcher, even after being told that he or she should find another job, would make no attempt to leave and would continue to operate as before. In such cases, the researcher's manager would be forced to initiate the employee termination process, an extremely torturous process for the manager. The process was long and cumbersome and involved setting up a performance improvement plan, holding periodic performance reviews with the researcher, requiring much documentation, and possibly taking over a year. Consequently, managers hoped that an underperforming researcher would have enough pride to leave Bell Labs on his or her own accord or when told to leave.

The life of the underperformers started to become difficult after Netravali took charge of Research. Anxious to improve the caliber of the researchers, Netravali urged the managers to identify underperformers every year, the bottom few percent of the researchers, and ask them to leave Bell Labs (leave Lucent or go to other parts of Lucent). As a result, Bell Labs managers finally started identifying the weak researchers and asking them to leave or be terminated. The process of terminating employees also became less tedious.

TOP OF THE CLASS

The Bell Labs Research organization that put Bell Labs and AT&T on the map in the world of computer science was the Computing Sciences Research Center. Within Bell Labs this center is also known as center 1127 or just 1127, its organization number.[70] The research contributions of 1127, particularly the creation of the UNIX system, and the C and C++ programming languages had an extremely significant impact within and without AT&T. Some other important contributions include various UNIX software tools, the Datakit switch, the computer chess champion Belle, and the recently developed OCELOT[TM] network planning tool for determining locations of cell phone towers.

When I joined Bell Labs in 1978, it did not take me long to figure out that within Bell Labs, 1127 was on top of the pecking order from the perspective of computer science or anything that had to do with software:[71]

Computing Sciences Research Center (1127)

Bell Labs Research minus center 1127
(I joined center 1135 when I transferred to Research in 1979)

Bell Labs development organizations

1127 has had many distinguished researchers such as Ken Thompson (retired from Bell Labs in 2001) and Dennis Ritchie, who together received the 1983 Turing Award for the UNIX system.

Researchers in 1127 had strong opinions about what they considered good software research. Because of the success of UNIX and C, many in Bell Labs accepted the opinions of 1127 as the gospel. Starting with the late 1970s, almost all new projects used UNIX and C. As a result, 1127 wielded tremendous influence within Bell Labs (including in the development organizations that were moved to the business units in 1984).

Soon after I joined Bell Labs, I collaborated with a colleague, Alan Feuer, to write a paper comparing the C and Pascal programming languages.[72] The thesis of the comparison was:

> *Pascal programs tend to be more reliable than C programs because of Pascal's strong typing. They also, because of Pascal's richer set of data types tend to be more readable and portable than C programs. However, C, because of its flexibility and lack of restrictions, can be used in a larger variety of programming domains than Pascal. ... C*

has been moving toward better program reliability and error detection through stronger typing.

This was around 1980. Since then, C has moved towards stronger typing as in ANSI C. Many C++ programmers, in the early days, used C++ only for its C subset to get the benefits of C++'s strong typing.

Our paper prompted Brian Kernighan, co-author of the famous C book, *The C Programming Language*,[73] to write a paper explaining why Pascal was not his favorite programming language.[74] He concludes by saying:

> *I feel that it is a mistake to use Pascal for anything much beyond its original target [of teaching]. In its pure form, Pascal is a toy language, suitable for teaching but not for real programming.*

Kernighan was right about standard Pascal. Over the next few years, Pascal lost to C as the programming language of choice. C won because of its flexibility and because it was riding on the coattails of the increasingly popular UNIX system.

However, C's lack of some facilities to support the writing of reliable programs was a serious issue. For example, C's lack of "array bounds checking" can lead to unreliable programs because C programs, typically inadvertently, can refer to inappropriate parts of computer memory. Some C aficionados claimed that array bounds checking was good for teaching, but not for production systems because it slowed programs and since C was used for writing "real" programs as opposed to toy programs, it did not need array bounds checking. Expert C programmers, they said, did not need such a crutch.

Unethical hackers have used the lack of array bounds checking in C to compromise some widely used C programs by forcing "buffer overflow" to get unauthorized access into networks and websites. They have on some occasions caused serious damage. Two decades ago, computer security was not much of an issue since most programs were written for a controlled user population that was usually small. With the Internet, the scenario has changed and programs such as Web servers, most of which up to now have been written in C or its derivatives, are used by large numbers of users, in some cases millions, on a daily basis. Corporate networks and websites are exposed to attacks from all over the world.

Our C-Pascal comparison paper had pointed out the strengths and weaknesses of both Pascal and C. The fact that I, a Bell Labs researcher, believed that C had some weaknesses and said so publicly put me on the wrong side of some in 1127. For example, I would occasionally get feedback from my managers that some researcher or some manager in 1127 did not think much about my work.

The C-Pascal comparison paper would haunt me for many years. Around 1984, I wrote a book about C[75] with the goal of teaching disciplined programming in C. Here is a quote from the book's preface:

> *C is a flexible programming language that gives a great deal of freedom to the programmer. This freedom is the source of much of its expressive power ... However, undisciplined use of this freedom can lead to errors.*

Before the book could be published, I had to get the approval of the Bell Labs Book Review Board[76] whose charter was to ensure that the quality of Bell Labs books was worthy of Bell Labs.

The Book Review Board consisted of representatives from Research and several development organizations in Bell Labs, about eight organizations in all. A few weeks after I submitted my C book for approval, the Book Review Board informed me that they would not approve the publication of the book. I was devastated. I had invested a tremendous amount of before and after work hours writing the book. I could not understand why the Book Review Board was not approving the book especially since I had received good feedback from my colleagues.

On further discussions with the Book Review Board, I learned that the Research reviewer had trashed the book and that the reviewer was from 1127. Each reviewing organization had one vote, but one no vote amounted to a veto. The Book Review Board would not give me the name of the actual reviewer, but the Research representative was Vic Vyssotsky, a senior executive, and 1127 was one of his organizations. Vyssotsky stood by his reviewer's evaluation and it seemed that I would not be able to publish the book, at least not with a Bell Labs affiliation and possibly not as a Bell Labs employee.

Fortunately, in the next few days, I was able to get all the reviewer comments from the Book Review Board. With the exception of the 1127 reviewer, all the other reviewers were positive about the book, with some being very positive. The 1127 reviewer did not give specific reasons explaining why he did not like the book and was, instead, very general in his negative remarks.

With the positive reviews in my hand, I confronted Vyssotsky. In light of the fact that every other reviewer was positive, Vyssotsky overrode the 1127 reviewer and approved the book. The Book Review Board then approved the book, *C: An Advanced Introduction*, for publication. Computer Science Press published my C book in 1985. This book received very favorable reviews in the marketplace and several versions of it were published.

Around this time, Bill Roome and I had started developing Concurrent C.[77] to be precise Concurrent C/C++. This programming language provided facilities for parallel programming, a model of programming particularly suitable for writing complex applications such as operating systems, banking applications, and control systems that need to handle simultaneously occurring events. Concurrent C/C++ was distributed to about 600 universities and research institutions. However, for Concurrent C/C++ to be used in serious projects within AT&T, it needed to be supported by a development organization. We also wanted AT&T to market it on a commercial basis.

1127 had very strong influence particularly in UNIX and C related matters. Since Concurrent C/C++ was not coming out of 1127, we from outside 1127 working on extending C and C++ had a hard time trying to convince a development organization to make Concurrent C/C++ into a product. Had we been able to get 1127's blessing, our task would have been easier. We tried, but the 1127 researchers we talked to had their own ideas about parallel programming with respect to C and C++. Although Concurrent C/C++ was distributed by AT&T as part of a collection of software tools, it was never made into a product partly because of lack of 1127's blessing and partly because AT&T had yet to establish a full-fledged software channel.

4 Looking For Dung But Finding Gold

B ELL LABS RESEARCHERS work hard and are very passionate about their research. However, they have a sense of humor and they occasionally come up with interesting and unusual ways of looking at things other than research topics. Bell Labs was known around the world as a great research lab, but the researchers also made it into a fun place to work.

PIGEON DROPPINGS

Arno Penzias and Bob Wilson discovered radiation from the Big Bang (their Nobel Prize work) while cataloging known sources of radiation. They detected radio static that was present regardless of where they pointed their horn-shaped antenna. Penzias and Wilson ruled out the possibility that the radio static was caused by man-made radiation. They did this by pointing the antenna towards New York City and other places on the horizon and finding that the radiation detected was not significantly more than the thermal temperature of the earth. They then checked the antenna and found no problem with it. Next, they speculated that the radio static was the result of the heat from the droppings of the pigeons roosting in the antenna. Getting rid of the pigeons and cleaning the antenna still did not remove the static. It was only after this that they concluded that the static was the result of background radiation. [78]

Ivan Kaminow, one of Penzias' colleagues at Bell Labs, joked that Penzias was an unusually lucky person:

> *Arno Penzias and Bob Wilson were trying to find the source of excess noise in their antenna, where pigeons were roosting. They spent hours searching for and removing the pigeon dung. Still the noise remained, and was later identified with the Big Bang. Thus, they looked for dung but found gold, which is just opposite of the experience of most of us.* [79]

MISSING BOOKS – A PRACTICAL JOKE

In the late 1980s, before the days of the Internet and the Web, the magnificent and well-stocked Murray Hill library was a valuable resource for the researchers. The library is located in the front part of the Murray Hill complex. Our offices were located at the rear end of the complex, far from the library. From our offices, it would take about five minutes to go down a few flights of stairs and walk to the library. Often it would take longer because invariably one was bound to run into a familiar face and spend sometime chatting with him or her.

One of my research colleagues was Harry, a dedicated researcher with a good sense of humor, who was always willing to chat. Harry had the best "private" collection of computer science books in Murray Hill. Harry had taken good advantage of the Bell Labs perquisite allowing researchers to buy all the books they needed for their research. His office was a mini-library! Books were neatly and alphabetically arranged on the several parallel shelves that lined his office walls. Instead of wasting time going to the library, we would often borrow books from Harry.

Whenever Harry lent a book, he would put a note with the book title and the borrower's name in the empty space left behind on the shelf. This allowed Harry to know which book had been borrowed and by whom. Harry only needed to send an email to the borrower and the book would be back in his hands within a short time. Harry was very generous in letting us borrow his books. Even when he was not around, we could borrow a book provided we left a note with the appropriate information.

In the 1980s, our offices typically did not have locks. They were left unlocked and many researchers even left their doors wide open. Only offices of senior managers and secretaries had locks to keep confidential documents confidential. One morning, Harry walked into his office and found that a book that he needed urgently was missing. There was the telltale empty space on the shelf where the book should have been, but there was no "borrower" note. Someone had borrowed his book without leaving a note with the relevant details. Harry was very upset. He walked up and down the hall complaining loudly, to anyone who would care to listen, about the inconsiderate person who had taken one of his books without following the appropriate protocol. Harry sent an email reminding us of book borrowing etiquette and demanding the immediate return of his book. Sometime later that day, Harry's book was returned anonymously.

A few days after the missing book incident, Harry walked into his office and noticed a big gap on his bookshelf:

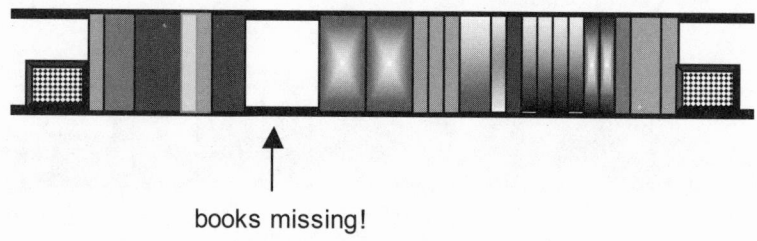

books missing!

Two or three books were missing and there was no "borrower" note. This time Harry was even more upset than before. He could be heard in the hall berating the unknown borrower or borrowers who had taken his books without his permission and without notifying him. He wanted a lock to be put on his door. He "flamed" all of us, the email expressing his anguish. He wanted the offender(s) to return his books immediately.

Unlike the previous time, the books did not show up on his desk by the end of the day. Harry was angry. In fact, several days went by without the books being returned. Harry continued to berate the unknown borrower(s) to anyone within hearing distance. He could be seen drinking the drink that soothed him – iced tea with lots of ice.

We were of course sympathetic towards Harry because, besides being a nice person, he was quite generous in letting us borrow his books. However, the missing books did not warrant so much anguish on his part. In the worst case, he could always order new copies (assuming he could determine what books were missing).

Then Morris, a colleague with a penchant for practical jokes and humor, collected a couple of us and said that he was going to show Harry where his missing books were. We followed him to Harry's office. He told Harry that his books were on the shelf, right in front of him. Harry was puzzled and so were we. There was a gap on his shelf – the books were missing, plain and simple!

Morris went to the bookshelf, put his hands on the two bookends, and pushed the books towards the middle. Now there was no gap. Morris announced with much fanfare, that Harry's missing books had been found!

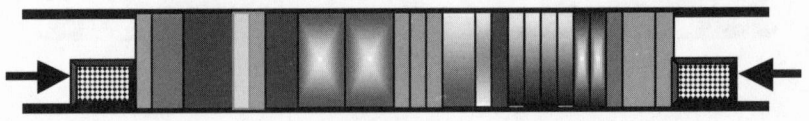

no books missing!

We all burst out laughing. Harry's missing books had never left his office. A strategic gap created by Morris had given the illusion of missing books.

PROVING (DISPROVING) THE EXISTENCE OF GOD

Bell Labs researchers were known to be supremely confident, at times arrogant, and sometimes a bit different. They often worked on esoteric research topics, with little or no relevance to the company, especially in the days when AT&T was a monopoly.

Shortly after yet another Bell Labs researcher won a Nobel Prize, the chairman & CEO of AT&T came to Bell Labs to congratulate the brilliant researcher, whom I will call Nelson. The AT&T chairman & CEO had a conversation with Nelson that went something like this:

> **Chairman:** Congratulations, Nelson! A well deserved honor!
>
> **Nelson:** Thank you, Mr. Chairman!
>
> **Chairman:** Nelson, now that you have achieved the ultimate, what are you working on?
>
> **Nelson:** I am counting the number of atoms in the universe.

The AT&T chairman was aware that Bell Labs researchers often worked on esoteric topics that were not related to the telecommunications business. The chairman continued:

> **Chairman:** This is very interesting. However, why are you doing this?
>
> **Nelson:** I am trying to prove or disprove the existence of God.

The AT&T chairman was taken aback and puzzled to hear that Nelson was trying to resolve an "open" theological problem, one that had been around from the beginning of the human race. What did God's existence have to do with AT&T? Why was Nelson spending time on theology? Bell Labs researchers were supposed to be working on technical topics. Maybe winning the Nobel Prize had affected Nelson. Nevertheless, the AT&T chairman continued:

Chairman: How can you do this?

Nelson: If the number of atoms in the universe is greater than a certain amount, then the universe will eventually stop expanding, it will start contracting, and then collapse back into the state before the Big Bang. This will be followed by another Big Bang, and the whole cycle will repeat itself. In such a scenario, the Big Bang is part of a natural cycle and there is no God.

If, on the other hand, the number of atoms in the universe is less than this amount, then the universe will keep expanding forever. In this case, some Force caused the Big Bang, and that Force is God.

Chairman: This is very impressive! However, what does it have to do with AT&T?

Nelson: God, like AT&T, is a monopoly. And I am studying monopolies.

This conversation is anecdotal and I heard it before 1984 when AT&T was still a monopoly. This banter may or may not have actually taken place.

SLEEPING AT MURRAY HILL TO AVOID COMMUTING

Many researchers spent long hours at Bell Labs. Researchers could be found at Murray Hill late at nights, on weekends, and on holidays. They would be trying to finish papers before conference deadlines, do research, debug systems, write software, etc. There were few distractions and interruptions after hours, which allowed the researchers to concentrate on what they really wanted to do.

However, not all the researchers who were at Murray Hill after hours were there because they were motivated by the desire to spend extra hours doing research. A few were there just because they were lonely at home. Occasionally, a researcher or two would be there so that they could impress their managers with their dedication to research. They would make sure that their managers heard about their working after hours. One researcher used to come after hours partly because he liked watching movies on the giant screen in our conference room, which was named the "Blue Flamingo Lounge." The Blue Flamingo Lounge had décor that matched its name and it was well known in Murray Hill as an example of researcher quirkiness.

In my view, a very dedicated and well-known researcher, who I will call Winston, takes the cake for his motivation to be at Murray Hill after hours. Winston, who was short and had a thick blond beard, was always around at Murray Hill. Apparently, he had too much on his plate, much of it his own doing. He worked long hours to catch up on his work. Some of us thought that he stayed around at Murray Hill for companionship since there was always someone to talk to, even late nights and on weekends.

One Friday evening, I saw Winston walking towards the Murray Hill train station, about ¾ of a mile from the Bell Labs Murray Hill complex. He was carrying a large and bulging red bag. The road did not have any sidewalks and Winston was walking on the side of the road, which was a bit dangerous. I stopped and asked him if I could give him a ride. He welcomed my offer and asked me if I could take him to the Summit train station, instead of the Murray Hill station. The Summit station has many more trains serving it since it is the junction of two NJ Transit lines, the Dover and the Gladstone lines, the latter being the one that serves Murray Hill.

The Summit train station was right on my way home and it was not a problem for me to take Winston there. As it turns out, I took Winston and his usually bulging red bag to the Summit train station several times over the next few years as a result of which I got to know him a little better.

Most of the time, when I went back to Murray Hill after dinner to work or when I went in on weekends, I would run into Winston in the coffee room or in the halls. I enjoyed chatting with him. He was very friendly and very knowledgeable about software, especially open source software.

I had heard rumors that Winston was occasionally staying nights at Murray Hill. One weekend I was chatting with him, when he confirmed that he often slept at Murray Hill, usually in a lab that had a sofa or in a conference room that had a comfortable place to sleep, and sometimes in his office.

Until very recently, there were no showers in the Murray Hill complex (at least none that could be used by the researchers). Therefore, I asked Winston what he did about taking a shower. Winston told me that he tried to go home every few days to clean up and put on some fresh clothes or, if he were unable to go home, he would go take a shower in a nearby gym.

As far as eating at night was concerned, Winston would buy extra food at lunchtime and heat it up in the microwave in our kitchen at dinnertime. Many organizations within Research had their own small kitchen with a coffee or a cappuccino machine, a microwave oven, etc. In addition, there were several food vending machines in Murray Hill. For weekend dinners, he would bring back some food from a nearby restaurant or the supermarket. The nearest restaurant and supermarket were about ¾ of a mile from Bell Labs (a little further than the Murray Hill train station).

One weekend, I talked to Winston about sleeping at Murray Hill. Here is how our conversation went:

Narain: Winston, why do you sleep at Murray Hill? Why don't you go home to sleep?

Winston: Sleeping at Murray Hill is convenient. I save a lot of time by not having to go home in the evening and then coming back in the morning. All I do when I go home is to eat, sleep, get up, and come back.

Narain: How long does it take you to go home?

Winston: A couple of hours.

Narain: How far do you live?

Winston: Not too far.

Narain: Then why does it take you a couple of hours to go home?

Winston: Well, I first have to walk to the Murray Hill train station, which takes me about 20 minutes. Then I have to wait about 5 minutes for the train to Summit. The train takes about 10 minutes to Summit. There I have to transfer to the Morristown line. Waiting for the Morristown train takes on the average about 20 minutes. The train ride then takes

```
about 20 minutes. Then I have to
take a bus. Waiting for the bus and
the bus ride take 30 minutes. Then I
have to walk 10 minutes.
```

Public transportation in most parts of New Jersey is rather poor especially if you want to go to some place other than New York City. A car is almost a necessity for working at Murray Hill. Winston should have been driving instead of taking public transportation. I wondered if he was trying to do his bit for energy conservation. Our conversation continued:

```
Narain: Why don't you drive home?

Winston: I don't know how to drive.

Narain: Then why don't you live closer to
        Bell Labs, say in New Providence?

Winston: New Providence is very expen-
         sive. I only pay a couple of hundred
         dollars for a huge apartment. I
         would have to pay three or four
         times that in New Providence.

Narain: Perhaps you could rent a smaller
        apartment. You are hardly ever in
        your huge apartment. By living close
        by you will save a lot of time and
        you will be able to relax at home
        and sleep in comfort.

Winston: I don't want to pay so much
         money.
```

I left it at that. I thought Winston was serious, but in retrospect, I think that he was humoring me. I do not believe that time and money were the real reasons why Winston was camping at Murray Hill. With his salary, Winston could easily afford an apartment close to Murray

Hill. Moreover, I had heard that he was also independently extremely wealthy. The irony is that Winston was financially so well off that he often forgot to cash his paychecks (he did not believe in direct deposits to the bank!). Winston had told me that every few months the company financial folks would call him and ask him to deposit his checks.

Winston was single. There was no one waiting for him at home. Although very friendly and helpful, Winston was a solitary person. I never did find the real reason for his sleeping nights at Murray Hill.

ELECTRICITY EXPENSES

Bell Labs was quite generous in reimbursing employees for expenses incurred by them for business reasons. However, some expenses were customarily not reimbursable. For example, I was told many years ago that "drinks" were not reimbursable even if they were incurred for a business reason. However, an employee could include drinks if they were part of a dinner. As a manager, when I was approving expense reimbursement forms, I would typically glance at them and then usually approve them. My assistant (secretary) would have previously scrutinized them for any mistakes. Most researchers were very careful about spending Bell Labs money. As a result, I rarely had a problem in approving the expense reimbursement forms of my researchers.

Attending to routine paperwork, such as approving employee expense reimbursement forms, was not my forte. I had made taking care of routine paperwork a weekly event in order to minimize the time I spent on it. My assistant would sit with me and explain items that I did not understand or did not care to understand.

Many Bell Labs researchers are among the brightest scientists in the world. Researcher Tom (my name for him) was one such person, well known in the scientific community. Tom would easily win the award for submitting the most creative request for expense reimbursement.

One afternoon, many years ago, my assistant brought over the employee expense reimbursement forms for me to look at and sign along with other administrative paperwork. I happened to take a close look at Tom's expense reimbursement forms since he had submitted two of them together and because the total amount was quite high. In one of his expense reimbursement forms, in the middle of the items listed for reimbursement, was an unusual item, one with a description that was something like

Home Electricity Use (6 months) $75

My assistant and I could not believe that an employee would ask to be reimbursed for the electricity that he or she had used at home. What was the electricity used for? Was it used to light up the home office when he was working late at night? For the home computer? I had never seen or heard of anything like this in my many years at Bell Labs. I found it hard to imagine anyone asking to be reimbursed for electricity. It had to be a mistake!

I did not approve Tom's employee expense reimbursement form with the electricity item and asked my assistant to return it to Tom. The next day Tom strode into my office demanding to know why I had not approved his expense reimbursement form:

Tom: Narain, why did you not approve my expense reimbursement form?

Narain: I didn't understand one of the items, the one about home electricity use. Did you really mean to voucher electricity used at home?

Tom: Yes.

Narain: What was the electricity used for?

Tom: The electricity was used to run the PC and the printer that I have at home to do my work.

Narain: Does it really cost $75 to run the computer for six months?

Tom: Yes, my calculations are based on the PC's power specifications.

Narain: In any case this request is very unusual. We do not reimburse electricity used for home computing equipment. Next thing I know is that you will be charging me for space used by your computer at home.

AT&T has paid for your home computing equipment, which I approved, primarily for your convenience. If you are not happy with the situation, then perhaps you should return the computer. If necessary, you should stay longer at work to use the office computer.

Tom: This is absurd! Electricity use should be reimbursed much like other items, such as travel.

Narain: Incidentally, in this case, since the PC generates heat, will you be willing to give AT&T a discount for the electricity cost since the PC reduced your heating bill?

Tom: Electricity use at home is a business expense!

With that comment, Tom walked away in a huff. I later realized that Tom could have easily countered my facetious argument that he give AT&T a discount for the heat generated by the PC. He could have argued that the lower heating cost in the winter would be offset by the higher air conditioning cost in the summer.

BETTER THAN BANKERS' HOURS

Officially, since the late 1980s, AT&T required employees to work a minimum of 40 hours. Sometime in the 1980s, the number of work hours had increased from 37.5 hours to 40 hours, with no corresponding increase in salary. The official AT&T work hours were from 8:15 AM to 5:15 PM with one hour for lunch.

Unlike in the business units, Bell Labs management did not expect its technical staff to generally follow AT&T's official work hours. Many researchers had unconventional working hours, but most were usually around at the office during the core hours, that is, some hours before lunch to some hours after lunch. Many researchers would come to office early and leave early or they might come late and leave late, or come at night and work all night. Moreover, many researchers worked many additional hours at home and would come to work on weekends. In essence, most researchers worked long hours, more than the 40 hours required by AT&T.

In the early 1990s, I had one programmer who was quite good technically. He was working with a couple of researchers, but their project was not moving at the speed I expected. After some investigation, I found that the programmer would come in around 9:00 AM and leave by 4:00 PM, his hours being better than bankers' hours (he also took some time off to eat lunch, usually at his desk). Moreover, when he left work, he left his work behind. Unlike most other employees, he would rarely, if ever, log on to his computer at work from home, even though I had allowed him to buy a top of the line PC for home use and was paying for his Internet connection.

Normally, as a manager, I did not care about the timings of my staff. Some employees preferred to work from home since there were fewer interruptions. However, the project was moving more slowly than I expected. I told the programmer that I expected more effort from him to ensure appropriate project speed. In particular, I told him that I expected him to work eight hours. Flex time, which required my approval, still required employees to work eight hours a day,

The programmer had been in Research for about a year. He would not believe my statement about AT&T expecting employees to work eight hours and that AT&T had official working hours. He said that it was a ruse on my part to make him work more hours. According to him, Bell Labs researchers and other employees came and went whenever they wanted and worked the number of hours they saw fit, and he did not think that they were required to work eight hours. I guess an employee observing the researchers' comings and goings superficially could come to such a conclusion.

To convince him about AT&T's formal work hours, I had to ask my assistant to show him the appropriate page from the AT&T GEI ("General Executive Instructions"), a very thick manual, with text in extremely small font, that contained all the rules and regulations that could possibly be of interest to employees. Grudgingly, the programmer started to work the expected eight hours. The project pace picked up a bit.

CONFERENCE PARADE

Because the pressure to publish was very high, Bell Labs researchers were very determined to publish papers. Publishing papers in conferences and journals requires a lot of effort and energy. For example, for a paper to be published in conference proceedings, the paper must first be accepted by the conference "program committee," which usually consists of a bunch of experts on the conference topic. A prospective author, responding to a conference's "call for papers," sends a copy (often multiple copies) of his or her paper to the program committee. The paper is then reviewed by some members of the program committee and rated according to some, possibly pre-specified, criteria. Some of the top papers are accepted for publication in the conference proceedings. The successful authors are notified of acceptance of their papers and given reviewer comments as feedback. They must then make changes, if any, suggested by the reviewers and submit a final version of the paper to the program committee. Authors whose papers are not accepted for publication in the conference proceedings are also notified and given reviewer comments.

Paper acceptance, rejection, and resubmission to another conference are a normal part of a researcher's life. The determination to publish papers was so high that if a paper were rejected, a few researchers would quickly "fix" the paper by incorporating the reviewer comments and then submit it to the next conference almost "overnight." Consider, for example, the following four most important "practical" database conferences.[80]

PRESTIGE	CONFER-ENCE	PAPER SUBMISSION DEADLINE	NOTIFICATION OF ACCEPTANCE/ REJECTION
Very High	SIGMOD	November	January
High	VLDB	February	May
Medium High	ICDE	June	September
Medium High	EDBT	October	December

Some researchers, both inside and outside Bell Labs, would start by submitting a paper to the most prestigious database conference, that is, SIGMOD. If the paper were rejected, they would be ready to submit it to the next most prestigious conference, VLDB, and so forth.

The paper submission and resubmission strategy could be described graphically as follows:

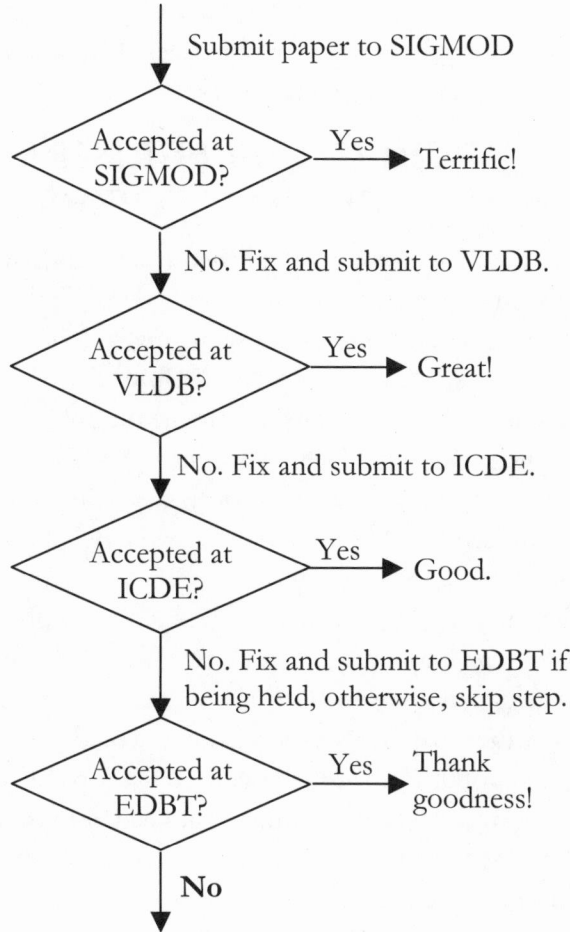

Submit paper to SIGMOD

Accepted at SIGMOD? — Yes → Terrific!

No. Fix and submit to VLDB.

Accepted at VLDB? — Yes → Great!

No. Fix and submit to ICDE.

Accepted at ICDE? — Yes → Good.

No. Fix and submit to EDBT if being held, otherwise, skip step.

Accepted at EDBT? — Yes → Thank goodness!

No

Give up or try again? Or send it to some other place?

Fortuitously for the database researchers, the conference dates for notification of rejection and paper submission deadlines gave the researchers enough time to do a quick fix on a rejected paper and then submit it to the next conference.

I am by no means implying that the above publication strategy was widely practiced. Many researchers submitted papers that had not been previously submitted to and rejected by the other conferences, directly to each of these conferences.

PUBLISH OR PERISH AND FEDERAL EXPRESS

The number of papers published is an important metric used by most major universities in determining whether or not a professor should be granted tenure. Consequently, junior professors aspiring for tenure must "publish or perish." Up to the early 1990s, Bell Labs culture encouraged researchers to focus on publishing as opposed to encouraging them to develop technologies for the business units. Many researchers published furiously as if they were coming up for tenure. Indeed, some researchers viewed Bell Labs as a temporary stop on their way to a university.

The intensity with which the researchers worked to publish papers was very impressive. A few days before a conference deadline, researchers could be seen working late nights at Murray Hill trying to complete their papers. Sometimes they would even stay the night at Bell Labs working to beat the conference's paper submission deadline. If, despite such efforts, they still needed more time to complete their papers, they would contact the conference program committee to request an extension.

After completing the papers, the researchers would send them overnight by Federal Express to the program committee. Federal Express picks up packages from Murray Hill at about 4:00 PM. If the researchers still needed a few more hours to complete their papers, they would take the packages to nearby towns that had Federal Express boxes with later pickup times. On occasion, the researchers would even take the packages to the Federal Express office at Newark Airport, about 16 miles away, where the packages could be dropped off as late as 10:00 PM.

For two or three weeks before conference deadlines, project meetings, internal presentations, (the few) interactions with business colleagues, etc., would be put on the back burner. Researchers with paper submission deadlines would balk at doing anything that would distract them from meeting the deadlines.

Publish or perish was an important part of the Bell Labs culture that was encouraged by management, many of whom had active research programs of their own. Only in the mid 1990s, did this culture start to change, moving towards working on topics to help the business units.

THE MORE THE MERRIER!

The value of basic or long-term research is often hard to measure objectively because its value only becomes obvious years after its completion when business applications, if any, become apparent. Even in the case of short-term research, the value is hard to determine unless and until the technologies and systems developed are deployed. Consequently, evaluating researcher contributions can be hard. One easy but not good performance measure used by some managers is the number of papers published. This led some researchers to focus on quantity rather than on quality.

A researcher publishing three or four papers a year in good conferences would do fine in performance evaluation, provided the papers reported significant research results. However, some researchers made publishing as many papers as possible their mission. One brilliant researcher, for example, had about 16 papers accepted for publication in one year. A few of them were in top notch conferences while the rest were in second- and third-rate conferences and workshops. To get 16 papers accepted, he probably wrote and submitted several more papers that were rejected. It was extremely unlikely that he was batting 1000 (perfect batting in baseball parlance) with such a large number of papers.

The papers written by this researcher were on several unrelated topics demonstrating a lack of focus. As his manager, I had a discus-

sion with the researcher, who I will call Tony, following his performance review:

> **Narain:** Tony, you have been extremely prolific this year. However, I do not understand one thing. Why would a researcher of your caliber and reputation submit papers to second- and third-rate conferences and work-shops?

> **Tony:** I have papers in all the prestig-ious conferences this year. However, I cannot expect all of my papers to be published in the prestigious con-ferences. Therefore, I was forced to send some papers to the weaker conferences and workshops.

> **Narain:** Our focus is on quality not quan-tity. I cannot believe that all your 16 papers report significant re-search results. It is important that you focus on significant innovation and invention and not on incremental results.

> **Tony:** All my papers do report significant results!

> **Narain:** I find this hard to believe since otherwise you would have been able to publish all your papers in the prestigious conferences.

> **Tony:** As you said, I have been prolific. I cannot help it if the prestigious conferences do not accept all my pa-pers.

Narain: I cannot believe that you can produce a significant research result every three weeks, especially in light of the other activities you are engaged in such as trying to build a prototype, traveling, and taking vacation, and on top of all this it takes time to write the papers.

After some discussion, Tony eventually agreed that not all his papers reported significant research results. Fortunately, over the coming months he changed his behavior and started to focus on quality.

PUBLISHING LOTTERY

Since some managers based performance on quantity rather than quality, several researchers devised strategies to publish as many papers as possible. One researcher treated conferences as a lottery. His philosophy was "the more papers you submit the more papers you will have accepted." He was right in observing that there was a significant element of chance in getting papers accepted in conferences. Besides quality, many other factors influenced paper acceptance or rejection. For example, factors such as the researcher's track record, the quality of the reviewers, their familiarity with the topic, and whether they liked or disliked the researcher[81] would all have some bearing on the acceptance or rejection of a paper.

Despite being capable of very good research, quantity, not quality, was at the top of this researcher's agenda. He would "crank out" papers quickly and submit them to conferences. Nevertheless, except for one year, he had a very good success rate in getting papers accepted in conferences. His manager knew little, if at all, about the researcher's area. We used to wonder about the quality of the program committees. Unfortunately, some young researchers looked at this researcher as a role model in building up a publication portfolio.

STOPOVER IN NEW JERSEY

Bell Labs was very generous in allowing researchers to attend conferences and would pay all their expenses for such trips. There was always money to go to reputable conferences, especially if the researcher was presenting a paper. A researcher in good standing could go to a conference at Bell Labs' expense even if he or she was not presenting a paper.

Most researchers traveled a reasonable amount, going to two or three conferences per year. However, a few researchers did take advantage of Bell Labs' generous travel policy to attend numerous conferences, workshops, and program committee meetings all around the world, often in exotic places. One colleague, for example, was attending so many such events that I would not see him for weeks at a time even though his office was just down the hall from mine. Once, when I saw him after an absence of several weeks, I jokingly asked him how he had been able to get a stopover in New Jersey. Ignoring my lighthearted remark, he proceeded to inform me that he would be leaving again to attend a conference in Asia in a couple of days. My colleague also informed me that he had accumulated a huge number of frequent flyer miles, which would allow him to travel around the world a few times without any approval from Bell Labs.

SPENDING ARM OF AT&T

Many people in the business parts of AT&T used to feel that Bell Labs was a cost center that spent AT&T's money freely, but did not do anything for the company business in return. Charles Wetherell, a friend from graduate school days at Cornell and a colleague at Bell Labs for some years, recently told me the following anecdote from the AT&T monopoly days. Bell Labs was then part of the big AT&T family, the Bell System, which included New Jersey Bell:

At Murray Hill, we had a New Jersey Bell employee whose only job was to take care of the phones and computer wiring in our build-

ing. One day, he was walking by my office giving a new New Jersey Bell employee the grand tour. As they passed by, I heard him explain, "This is Bell Labs, the spending arm of AT&T. It's a dirty job, but somebody's got to do it."

BELL LABS AND AT&T BRANDS

The Bell Labs brand is a global brand, known globally because of Bell Labs' reputation as an invention powerhouse, its famous scientists, the tens of thousands of awards won by its scientists, and because Bell Labs scientists interact with their counterparts all over the world. Bell Labs brand equity was the main reason why Lucent used the endorsement line "Bell Labs Innovation" in its logo.

The AT&T brand, especially in the monopoly days, although very well known and a highly trusted brand within the USA, was essentially unknown outside the USA. This was because for decades AT&T's business operations had been primarily limited to within the USA.[82] It was only after the 1984 breakup that AT&T started to globalize once again.

Shortly after the AT&T divestiture, I heard the following anecdote that highlights the difference between the global recognition of the AT&T and Bell Labs brands. I cannot vouch for its veracity, but it does illustrate the importance of the Bell Labs brand as a global brand.

An American trade delegation consisting of senior business executives, CEOs and the like, was visiting a very big Asian country, an emerging market with a huge business potential. The delegation included, among others, the AT&T chairman & CEO and the Bell Labs president. At the ceremony welcoming the delegation, the AT&T chairman & CEO was introduced to the Minister of Trade. The Minister went through the usual pleasantries and welcomed the AT&T CEO to his country. Then it was the turn of the Bell Labs president to be introduced to the Minister. This time the Minister's reaction was very different. He welcomed the Bell Labs president effusively as a much-honored guest, the head of the prestigious Bell Labs, a company that was very famous in his country. He praised Bell Labs' great achievements in science and technology. The Minister then requested the Bell Labs president to meet with him privately for dinner and business discussions.

The Minister was apparently not aware that Bell Labs was just a division of AT&T. He was also obviously not aware that the AT&T chairman & CEO standing nearby and watching the Bell Labs president being feted was the boss of the Bell Labs president. Incidentally, besides the AT&T chairman & CEO, there were several other people higher than the Bell Labs president in the AT&T corporate hierarchy.

Given the global name recognition of Bell Labs, it was therefore not surprising that Lucent made the Bell Labs name part of its commercial identity. It would take some time before the name Lucent was established as a brand both within the USA and globally. Lucent management correctly assumed that associating the Lucent name with Bell Labs would help build the Lucent brand faster and open more doors across the world.

5 Do We Work For The Same Company?

U P TO THE 1980s, most Bell Labs researchers had little or no contact with their AT&T business unit colleagues. Business unit executives rarely came to Murray Hill to tell the researchers about their businesses, products and services, customer needs, challenges, vision, business strategy, etc. Bell Labs had operated as a university-like institution for so long that it had become disconnected from the business units. Bell Labs was not working with the business units to develop new products and technologies.

The disconnect between Bell Labs and AT&T business units manifested itself not only at work, but also in social gatherings. Since AT&T was the biggest employer in New Jersey,[83] I would often meet AT&T business unit employees in social settings. In fact, in the 1980s, I met more people from the AT&T business units outside work than at work. To break the ice, we would make small talk, which would eventually veer around to talking about work. Having established that we worked for AT&T, we would find that we did not have much in common professionally. Our work was very different. Mine was to do research, and theirs was to develop, market, or sell products and services. From a technology perspective, there was no overlap in my research and their products and services since I specialized in computer science and they focused on telecommunication products and services.

One person I met was Satish Mehta who worked for one of the AT&T business units. In trying to place me organizationally within AT&T, Mehta would reel off names of important senior executives in AT&T (actually in his business unit). I would barely recognize a name or two. Similarly, Mehta would not recognize the names of the key executives in Bell Labs Research.

After a few such encounters, I came to realize that there was little, if any, interaction between Bell Labs Research and AT&T business units. It was as if Bell Labs Research and the AT&T business units belonged to two different companies, each marching to the beat of its own drummer!

After entering the competitive world, AT&T started expecting a return on investment from Bell Labs Research. Consequently, the business units and Bell Labs started the process of getting to know each other and working together. However, it took a few years after the divestiture for this process to get started.

TWO DIFFERENT WORLDS

Before Lucent located its headquarters at Murray Hill a Bell Labs researcher could go for years without meeting anyone from the business units. In the heyday of the AT&T monopoly and for some years after that, Bell Labs researchers would on a rare occasion go to AT&T's magnificent headquarters in bucolic Basking Ridge for a meeting or a presentation. The headquarters, which were located on a 140 acre campus, included AT&T's landmark building and its famous "Golden Boy" statue.

The gilded statue, named the "Spirit of Communications" by AT&T, was designed by Evelyn B. Longman and built in 1916. The 24-foot tall statue depicts a Greek god-like figure of a nude boy with wings. The Golden Boy statue stood for decades on top of AT&T's building at 195 Broadway in lower Manhattan before AT&T moved it to its building in midtown Manhattan and then to Basking Ridge.

In the Basking Ridge complex, Bell Labs researchers could be easily identified as the odd birds because of their attire. AT&T corporate executives typically wore blue or gray suits, white shirts, and ties. The visiting researchers would stand out since they would typically be wearing casual clothes such as T-shirts and jeans.

Since the researchers had little or no contact with the business units, most of them did not know much about the business units nor did they care to. They and their managers had few, if any, practical thoughts on how they could help the business units or work with them. On the other side of the coin, the business units knew little about Bell Labs Research, how it operated, or how they could leverage it to help their businesses. They did know, often from the media, that there were some very famous scientists at Murray Hill. However, the business units had little, if any, contact with Bell Labs Research, and, as a result, Bell Labs Research did not fit into their business equations. The business units therefore did not involve the researchers or their managers in planning strategy, defining the product roadmap, customer sales visits, etc.

Research involvement in business unit activities was necessary so that the researchers and their managers could understand business unit needs and prioritize their research projects. In the absence of such involvement, researchers picked research topics based on their interactions with the academic world through collaborations, technical conferences, and technical journals. Starting in the 1990s, researchers started to see business unit PowerPoint presentations on strategy passed down by senior Bell Labs management. Unfortunately, these presentations tended to be out of date because there was no formal process of getting this information to the researchers or their managers in a timely manner.

As we moved into the 1990s, attempts were made to engage Research with the business units. At first, researchers and their managers did not understand why the business units were not automatically interested in new research technologies and instead talked about customer needs, business justification, and their existing products. On the other hand, the business units did not understand that the researchers' role was to invent new technologies for the next generation of products and services. The researchers were not just a bunch of programmers who would, at the business units' beck and command, perform routine development tasks for them, tasks that could be done by their own employees.

THE WALL

Bell Labs Research was separated from the business units by a wall that hindered communication and collaboration and led to cultural differences. The wall was both organizational and physical. Bell Labs Research was not under the direct control of the business units. It was funded directly by the corporation out of central funds and the business units had no say in directing this funding to specific research projects. Until 1984, Bell Labs had been incorporated as a separate company, which ensured that the business units would have little, if any, influence on specifying Bell Labs research topics and direction. The rationale for keeping Bell Labs Research away from business unit control was to give the researchers freedom to work on next generation technologies, free from business unit pressures to work on short-term development projects.

Physically, most of Bell Labs Research was located at the Bell Labs headquarters in Murray Hill away from the business units and their day-to-day pressures. A small part of Research was co-located with various business units in Holmdel, NJ and later on, a score or so researchers were co-located in Naperville, IL with the switching business unit. The co-located Research organizations did have more interaction with the business units than the Murray Hill researchers.

A chasm existed between Research and the business units. While the business units felt that research projects did not have much to do with the company business, the researchers believed that their work was important for the long-term prospects of the company. The business units believed that the researchers had little or no understanding about the business value of their work, but also that the researchers expected the business units to be eager to step up to the plate, take their research gems and make them into products.

It was the organizational and physical separation of Research and the business units that caused this chasm and the development of a cultural gap that manifested itself as ignorance about each other and a mismatch of expectations. For example, researchers would develop a prototype system and attempt to "throw it over the wall" to the business units. They believed that they had done the hard work of developing new technology, and that it was now up to the business units to make the prototype into a product and take it to the market.

On the other side, the business units did not trust the researchers or their prototype systems. They did not believe that the researchers were committed to working with them to help make the prototypes into real products or to provide user support. As a result, the business units wanted the researchers to deliver products that were better than anything available in the market, had customers, and were supported by organizations other than Research. These expectations were generally a showstopper since Research was not set up to do what the business units wanted.

The wall separating Bell Labs and business units is being torn down. Making Bell Labs a division of AT&T started the process of eliminating organizational barriers between Bell Labs and the business units. Locating Lucent's headquarters at Murray Hill initiated the process of reducing the physical barriers between them.

CULTURAL DIFFERENCES

For most of its nearly 80-year history, Bell Labs Research was kept at arms length from the rest of the company. The separation was strongest in the AT&T monopoly days and, over several decades, it fostered a university-like research culture in Bell Labs Research. This culture prevailed through the 1980s and lingered on into the 1990s. Because of decades of freedom from business constraints, Research evolved into an ivory tower where researchers worked on topics of their own choosing, ignoring "minor" issues such as business relevance.

Bell Labs traditions and culture were very different from those of the rest of the company. Just as a successful business venture in a foreign country requires understanding the country's traditions and culture, Research and the business units needed to understand each other's traditions and culture to work together successfully. Until recently, there was no motivation or necessity for either side to invest the extra time and effort required to lay the groundwork for successful collaboration.

For starters, the motivation of the researchers was not the same as that of their business unit counterparts because of the different value and reward systems in Research and the business units. Researchers were motivated by inventing something new, publishing papers, being on conference program committees, and establishing a professional reputation. Researchers were rewarded for their success in these academic pursuits and there was little or no pressure to correlate researcher contributions with business needs.

Researchers operated in the world of "curiosity-driven" research that would establish them in the academic community to which some of them hoped to migrate at some point in their careers. In the late 1990s, in the era of the startups, founding or joining a new venture that held the potential of quick riches replaced the goal of going to a university. Following the Lucent spinoff in 1996, in which Lucent positioned itself as a Silicon Valley type company, there was great urgency and pressure on the Lucent business units to deliver new products and do so quickly and to generate increasingly higher revenues. Furthermore, Wall Street expectations and Lucent's stock valuation required Lucent to grow revenues at a very high rate, which was more appropriate for small companies or startups. For the first time, the urgency to deliver business value started percolating down to the researchers, who had until then been living in a world insulated from business issues. For example, many managers started giving researchers working on projects with business value more resources and rewards, such as programmers and stock options, and they started to publicly praise such projects. They also looked to hire new researchers with system building skills as opposed to paper publishing skills.

Raises and promotions were not a big motivating factor for many researchers. Most of them put a larger premium on the freedom to work on topics of their own choosing than on getting a larger raise – some of the researchers were quite well paid. Many of the outstanding researchers did not aspire to be promoted because they were more influential than many managers and they preferred a life without management headaches.

In contrast to the researchers, business unit staffers operated in a world of revenue and profit, and their goals, which were usually short term in nature, were well defined. Unlike researchers with their long-term research goals and no short-term pressures, the business unit staffers were under constant pressure to meet or exceed, their yearly targets. The pressure to deliver on target started in earnest after the AT&T divestiture that, with the stroke of a pen, put AT&T in the midst of competition.[84] The pressure to deliver increased every year as AT&T kept losing market share to its competition even though its overall revenues were increasing slowly or holding steady.

Measuring researcher contributions objectively has traditionally been a difficult task. It is hard to put a value on research contributions in terms of dollars and cents, at least in the short term. In contrast to researcher contributions, objectively measuring the contributions of the business unit staffers is relatively easy. Unlike research deliverables, business unit deliverables are typically more tangible, for example, they can be making business or product plans, developing systems, meeting product delivery dates, or meeting sales targets. Raises, bonuses, and chances of promotions of the business unit staffers depend upon whether or not they meet their deliverables. A promotion is very important in the business units, because not being promoted to a manager, over time, reflects poorly on the employee.

Research and the business units had very different work styles. Research had a *laissez-faire* work environment with researchers being free to set their research directions and select their projects. In contrast, management direction and guidance played a strong role in assigning tasks and projects to the business unit staffers. Research projects either had no deadlines or they had soft deadlines set by the researchers themselves. Researchers were not used to projects with hard deadlines (except publication deadlines). Business unit projects, on the other hand, usually had deadlines that had to be met. Most researchers seemingly worked at a leisurely pace, except a few times a year, when they had to meet publication deadlines. In the business units, crunch periods came when new product releases or customer deadlines were in the offing. Finally, many researchers often worked on multiple projects simultaneously to hedge their bets. They also had other commitments such as being on conference program committees and editorial boards. In contrast, business unit staffers usually focused on a single project.

The systems produced by Research and the business units were generally very different in nature and built for different reasons. Research systems, which were generally prototypes, were meant to be "proof of concept" systems. They were not built for real users and customers, and they therefore lacked many necessary attributes of products such as good user interfaces and documentation. Business unit systems, built to be products were developed using focus groups, concept demos, customer meetings, code reviews, etc. Most researchers did not have the skills, experience, and resources to build real commercial quality systems.

The researchers and the business unit staffers even dressed and communicated differently. The researchers dressed informally, T-shirts being a common part of their attire (most researchers were men). Business unit staffers tended to dress more formally, especially when meeting customers. Most researchers preferred informal discussions to formal meetings. In contrast, formal meetings in business units were an essential part of the job. Many researchers preferred email while many business unit staffers preferred voicemail.

Collectively, these differences made collaboration between Research and the business units challenging.

RESEARCH'S VIEW OF BUSINESS UNITS

Researchers, based on their interactions with the business units, did not have a high opinion of the technical expertise of the business units, especially their understanding of and interest in technology trends and new technologies.

The researchers felt that the business units were not interested in working with Research. The business units did not involve the researchers or their managers in developing their technology strategy and product roadmaps. Moreover, information about their strategy and product roadmaps was not easily available to the researchers.

The researchers believed that the business units paid only lip service to new technologies, especially in front of senior Research leaders. Business units were not seriously interested in new technologies. They just wanted to sell more of their existing products not realizing that other companies were developing new technologies that would "eat their lunch." Many business unit managers did not appreciate new technologies and the positive impact they would have on their future revenues. They were driven by short-term business needs and, consequently, they focused on incremental product enhancements.

Business units were technology followers not technology leaders. Before adopting a new technology or commercializing a research system, they wanted market research, proof of demand, sales estimates, etc. Unfortunately, most research technologies and systems were too far ahead of the market for such information to be found and, if found, for it to be useful. The business units were more comfortable with technologies that had already entered the market rather than with championing new technologies.

Business units preferred to acquire new technologies and products, instead of developing them jointly with Bell Labs.[85] They wanted Bell Labs to give them products that they could take to the market.

BUSINESS UNITS' VIEW OF RESEARCH

The business units were not interested in most Bell Labs research technologies and projects because they did not address their business needs. They felt that Bell Labs researchers and managers did not understand or care about what the business units wanted or needed. And, instead of working with the business units to understand their current and future technology needs, the researchers preferred to throw half-baked research technologies over the wall to the business units. Because of many unproductive dealings with Research over the past few decades, the business units found it more effective to operate without Research. Given their experiences with Research and their perception of the researchers, the business units were reluctant to work with Research and this reluctance seemed rational.

Business unit staffers felt that Murray Hill researchers lived in an ivory tower, disconnected from the rest of the company and from the business world. The corporation had put Research on a pedestal. The researchers were encouraged to think and behave as privileged employees by being given special perquisites such as private offices, higher salaries, and the freedom to travel to several conferences, which were sometimes held in far away places that were known more for tourism than for science or technology. And on top of this, most research activities did not have much to do with the business of the company.

The business unit staffers felt that the researchers were arrogant and had a low opinion of the business units. In particular, the researchers behaved as if those in the business units were not bright technically and that they did not know or understand the latest technologies. Researchers did not appreciate that the business unit staffers did not have the time or luxury to dabble in every new technology that came along. They had to generate revenues, some of which went to support the researchers.

The business unit staffers viewed the researchers as defocused and lacking in commitment. The researchers worked simultaneously on several different projects, had too many external commitments, and did not view collaborations with the business units to be a critical part of their job. There was no penalty for the researchers to walk away from collaborations with the business units. Research management did not guide the researchers to focus on business value. The researchers worked on whatever they wanted, whenever they wanted, and with whomever they wanted. They were simply not aligned with business unit goals.

Researchers did not seem to understand that business units had to make money. Just because a researcher thought that a technology was interesting was not enough justification for the business unit to invest resources in developing the technology. Business units needed to justify their investments with business cases.

Researchers built prototypes to validate their research ideas and then wanted the business units to take the prototypes and make them into products, and then build businesses around them. The business units believed that the researchers did not have a clue about the business potential and value of their work, what it took to make a product, market it, sell it, and provide customer support. From the researcher's perspective, once a prototype had been built, the hard part of developing a new product had been done. The business units believed the researchers viewed taking a prototype and making it into a product as mundane work that was beneath their dignity and a waste of their energies.

Finally, the business units viewed the researchers as prima donnas who were hard to work with and manage, and who exhibited no serious commitment to working with business units. The researchers did not want to work on projects and topics that addressed business unit needs. Instead, they simply wanted to peddle their pet research projects.

BELL LABS & THE BUSINESS UNITS

Despite the fact that corporate leaders had put Bell Labs Research on a pedestal, the business units, by their actions, made it abundantly clear that they were not enamored with Research. Business unit complaints that the researchers were not addressing their needs represented one part of the classic "chicken and egg" story. The researchers were more likely to address business unit needs if they understood what they were. To remedy the situation, the business units needed to involve Research in their strategy planning and roadmap processes.

The corporate leaders did not foster a culture that would make Research and the business units work together. There was not much reason to do so as long as AT&T was a monopoly. When they started prodding the researchers and the business units to work together, they encountered resistance. The researchers wanted to work on science. On the other hand, the business units were reluctant to work with Research because they felt that their leaders were forcing them to collaborate with Research on projects that did not make business sense. Projects initiated by Research were not on the business unit roadmaps, did not have a big market potential, would compete with existing business unit projects, or would defocus the business units. Moreover, to work with Research, the business units would have to divert resources away from existing projects since senior management would not give them additional resources. The business unit staffers also resented the researchers, the Johnnies-come-lately, for claiming to be experts in understanding business unit needs and issues. After all, the business unit staffers had been in the game much longer than the researchers, most of whom had never even talked to a real customer.

Left to their own devices, the business units would have preferred to not deal with Bell Labs Research. For example, one product manager I talked to said that his performance was measured on short-term goals. If Research got involved in his product and its evolution, the product manager was afraid that he would lose control. Researchers were not committed to business unit issues, they did not appreciate that a product manager has to make money, and that the business units do not dabble in new technologies just because they are interesting. If he, as the product manager, disagreed with the researchers, then all hell would break loose because the researchers' view of the disagreement would quickly reach his business unit leadership. Researchers had direct connections to the top Research leaders, while he, because the business units were hierarchical, had to go through several levels of management before reaching the top leadership. The researchers would complain directly to the top Research leaders, who in turn would complain to his business unit leaders, and then he would be hauled up to explain the issues. It was just simpler to not deal with Research.

For Research and the business units to work together successfully, there needed to be a desire to do so on both sides. The business units needed to involve Research to work with them to develop new technologies and products, that is, they needed to "pull" Research to work with them. On the other hand, Research needed to convince the business units about the value of its new technologies and its desire to work with them. That is, Research needed to "push" its technologies into the business units. Unfortunately, there was little pull from the business units and, until the early 1990s, there was little "push" from Research. In the 1990s, Research leaders, particularly Arun Netravali, determined that the long-term health and even survival of Research depended upon producing new technologies and products for the business units. As a result, Research led by Netravali tried to proactively push next generation technologies to the business units.

Netravali generated this push by "beating" upon the researchers and their managers. In the 1990s, the Research push was not matched by a corresponding pull from the business units. For effective collaboration between Research and the business units, the push from Research had to be matched by a pull from the business units.

This lack of pull from the business units had been a source of a great deal of frustration for the researchers who wanted to help the business units. According to Dennis Ritchie,[86] author of the C programming language and co-author of the UNIX system,

> *For some of us, in fact, a principal frustration has been the inability to convince others [AT&T business colleagues] that our research products can indeed be useful.*

For researchers developing software systems, as opposed to hardware, dealing with the business units was especially frustrating. The business units were comfortable selling physical "boxes" such as switches but not software. They were more amenable to working with Research on next generation systems and technologies that could be sold as boxes. For example, the optical networking group[87] (ONG) collaborated with the physical sciences division of Research to make their LambdaRouter™ all-optical cross-connect switch, which uses MEMS (Micro Electro Mechanical Systems) based mirrors, into a product. The mirrors, which are tiny, are used to reflect, i.e., switch, laser beams from one fiber to another. Without Research, ONG could not have built the LambdaRouter because they lacked the requisite expertise with MEMS based mirrors, a key technology needed for the LambdaRouter.

The business units did not know how to make money by selling software. Most of their software systems were bundled in the boxes sold by them. The business units also did not appreciate software research because they felt that they could always extend their existing systems with new capabilities. For example, instead of offering a Web-based customer care system as a product, the Lucent BCS business unit (now Avaya) preferred to extend their existing PBX with Internet capabilities. A Web-based customer care system, a pure software system, would sell for much less than the enhanced legacy system, which came with proprietary hardware. Marketing of a pure software-based customer care system was consequently viewed negatively, even though such systems were being sold by companies such as Cisco and would compete with Lucent's legacy PBX systems.

To get a taste of the challenges researchers had to deal with in establishing joint projects with the business units, consider the following two research projects, one on real-time billing and the other on Web-based customer care. These projects represented next generation technologies that were potentially important for the business units. In both cases, the business units initially did not want to work with Research, but our persistence led to collaborations between the business units and Research.

TOO BUSY TO WORK WITH RESEARCH

The goal of the Sunrise project, as mentioned earlier, was to help AT&T in the area of billing, a topic that was a far cry from academic research. This was clear evidence that Bell Labs was changing to align itself with business unit needs.

It was my responsibility, as the manager, to help the billing researchers partner with the business units. One potential partner was the AT&T Business Communications Services (BCS) business unit,[88] which provided long distance service to businesses. I had met a vice president from the BCS billing systems organization at a conference several years ago. The odds of my meeting AT&T business unit executives at the kinds of technical conferences the researchers attended were close to zero, but we were both there because a business school had organized this conference. I went to visit the BCS vice president along with two of the researchers. We explained to him the motivation behind Sunrise and described its capabilities. I told him that we would like to work with BCS on real-time billing. The BCS vice president said that he was very busy, had many commitments, and was not interested in working with Research.

We encountered similar resistance from others in BCS, which was under a lot of pressure because of competition. As a result, most BCS folks that we talked to did not have the time, the resources, or the desire to work with us. Nevertheless, we persisted in trying to find a partner in BCS. Although it took us many months, our persistence paid off, and we established a joint project with BCS on real-time billing. Involvement of the senior leadership of Research and BCS had been necessary to establish the collaboration. Eventually we also established a partnership with the AT&T Consumer Communications Services (CCS) business unit. These two partnerships represented a big success for Bell Labs in its efforts to help the business units.

Go Work On Something Else

Late in 1995, following the billing project, I decided to explore the area of customer care. AT&T was spending a substantial portion of its revenues on customer care and therefore, reducing customer care costs was an important issue for AT&T. To better understand customer care, I visited several groups in the various business units. Even though AT&T was complaining about high customer care costs, the business units did not ask Research to assist them in reducing costs. Instead, it was Research that took the initiative to explore issues in customer care.

After several unproductive attempts, we connected with an AT&T BCS group that was trying to understand how they could give their customers a single point of contact, instead of many, for customer support. Their organization offered vertical industry solutions to customers by combining products and services from different parts of AT&T. For example, in their customer care solution for the insurance industry, the computers would come from NCR,[89] 800 number service from BCS, email from AT&T Easylink Services, and software from third parties. Customers would get multiple bills and they had to deal with multiple business units for customer support. It was obvious to the customers that the AT&T product offering was not one product, but an aggregation of several products. BCS was not able to present a unified solution to its customers because it had to deal with other AT&T organizations that had their own independent databases, applications, and processes. To enable BCS to present a unified solution, we proposed a system based on innovative Web and database technology.

Unfortunately, even before we could seriously engage our BCS counterparts, the AT&T trivestiture was announced in 1995. Bell Labs and BCS would be in two different companies. BCS would go to AT&T and Bell Labs would go to Lucent. This announcement quickly killed our customer care technology interactions with BCS.

Fortunately, our customer care research led us to propose a new type of call center based on the Web. This Web-based call center (customer care system) would be like a traditional telephone call center except that customer and call center agent interactions would be Web based and would initially use text chat instead of telephony. Later on, we would incorporate voice chat and following that telephony based on the Internet technology called VoIP (Voice over the Internet Protocol).

We discussed the Web-based call center with the Lucent call center folks (Lucent BCS[90]) in Denver. We described our vision and told them about our technology. Many in BCS were interested in our ideas, but establishing a joint project with BCS was going to take time and persuasion. For example, one BCS call center development manager suggested that we would be better off working on a different topic, such as workflow, instead of call centers. The manager would not give a rationale for suggesting workflow except to say that workflow was an interesting research topic. We got the feeling that this was a turf problem. BCS had been working on call centers for many years. Research, on the other hand, was just entering the call center arena. Nevertheless, we were persistent. Netravali, then vice president of research, was able to establish a partnership with BCS by using his connections with the BCS leaders. Research and BCS were together going to build a Web-based call center.

The key concept behind the Web-based call center was the notion of shared browsing, also known as co-browsing, which allowed two or more users to see the same Web page on their browsers. If one user moved the browser to another page, the browsers of all the other users would move to the same page. Vinod Anupam and I had invented the concept of shared browsing on the Web and we had applied for patents.

In 1997, the BCS folks in Denver took some of the code from our Web-based call center prototype and integrated it with their traditional PBX-based call center. This Web-enabled call center, which was called the Internet Call Center, won several prizes and awards. For example, the Internet Call Center won the Bell Labs President's Silver Medal for Innovation & Technical Excellence in 1998 and it was the named the product of the year by the Call Center Magazine, also in 1998.

We continued our joint efforts on building a pure Web-based call center, which we named Stair9, an unusual name for a Lucent offering. Our product manager proposed the name, inspired by the fact that the stairway leading up to our offices in Murray Hill was named Stair 9. By late 1999, Stair9, or NetAssist as it was later called, was built and ready to go. At this point BCS essentially abandoned

NetAssist saying that they did not have the channels to sell software and because it could hurt the sales of the voice based call center, which included their PBX offering that was generating good sales. BCS had never been enthusiastic or convinced about NetAssist. They had been nudged into partnering with Research without having done the necessary market research and business case analysis. Had they done their homework and then decided to work with Research, in all likelihood they would have marketed NetAssist.

We, as researchers, had done our job. Our research had helped BCS build the prize-winning Internet Call Center and thus be on the forefront of call center technology. However, we were disappointed that the pure Web-based call center did not make it to the marketplace. In baseball terms, we had hit a single, but not the home run that we had really wanted.

POSTSCRIPT

I strongly believe that researchers working on challenging problems, even if they are related to products, can find interesting research results. For example, while implementing co-browsing, in the summer of 1997, Anupam found a security loophole in JavaScript. Despite using https (the secure version of the Internet protocol http) and encrypted passwords, JavaScript could be used to capture the text entered on a Web page and send it to a third party website. Anupam worked with Netscape and Microsoft to help close the security loophole. Discovering this loophole generated a lot of publicity for Anupam and Bell Labs. Anupam's discovery, for example, was mentioned in media outlets such as CNET.com, CNET Radio, Newark Star-Ledger, Bloomberg Business News, Seattle Post-Intelligencer, and so forth. Anupam also co-authored a paper on browser security.

While working on customer care, we had come up with the idea of co-browsing. We had been the first to invent this technology and we had patented it. Many companies, including SMALL, were using co-browsing technology in their Web-based customer care systems.

However, Lucent was not benefiting from our patents on co-browsing. In 1999, I suggested to our patent lawyer that it was time for us to sue companies such as SMALL for violating our patents, if they were indeed doing so. After some investigation, the lawyer told us that SMALL was a startup (which we knew) with little revenues and it was not appropriate for Lucent to pursue them for patent infringement at this time. If Lucent sued SMALL, the smaller company would benefit significantly in the public relations arena when the news got out that a giant company like Lucent was suing SMALL. He suggested that we wait until SMALL started generating significant revenues.

Some months later, in September 1999, BIG bought SMALL for several hundred million dollars. BIG was a giant company like Lucent and had huge revenues. Now was the time to enforce the co-browsing patent. I went back to the lawyer and updated him on the acquisition of SMALL by BIG. Perhaps, we could now sue BIG for patent infringement. The lawyer informed me that Lucent and BIG cross-licensed patent bundles to each other. Unfortunately, the co-browsing patent was included in the bundle Lucent had cross-licensed to BIG and therefore it was not possible to sue BIG.

Finally, after BCS decided not to market Stair9, Anupam and I decided to take the co-browsing technology to the marketplace directly. We would build a website that businesses could use to quickly enable their sites with Web-based customer care capabilities. We would thus be an application service provider for customer care.

Using the technology we developed for Stair9, we built the Surf N Chat website:

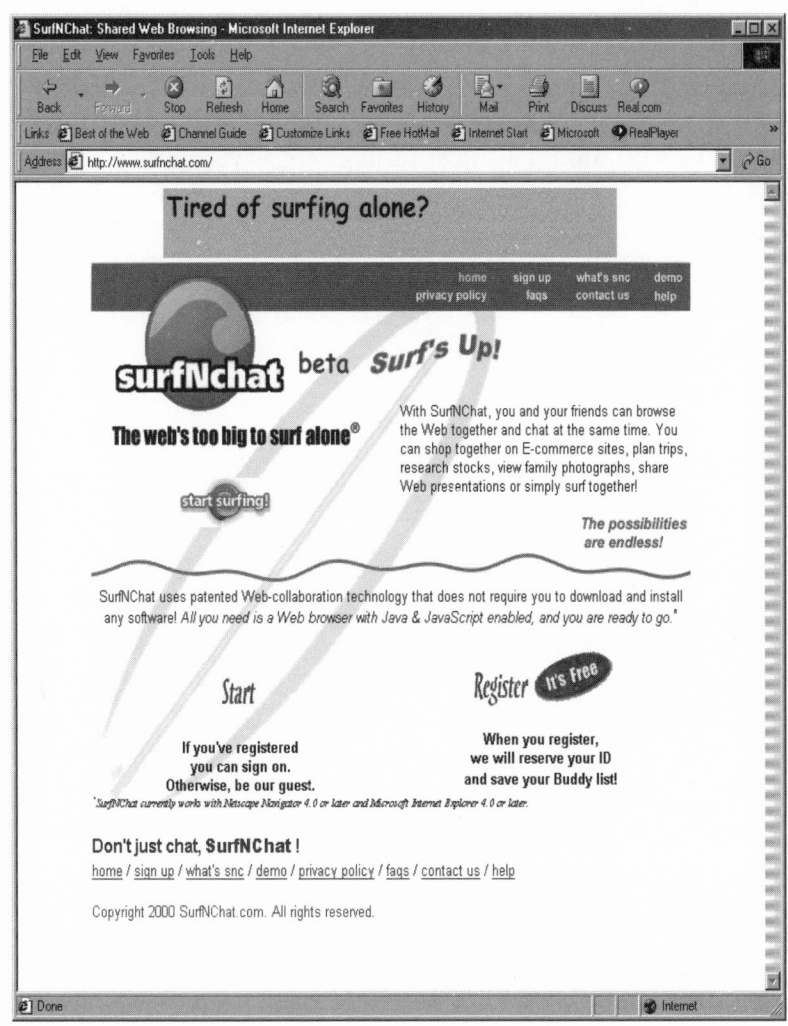

Besides offering customer care facilities, Surf N Chat also allowed users to do what its name says, i.e., surf and chat together at the same time. We tried to spin off a new venture via Lucent's New Ventures Group. Unfortunately, by this time, the market had started to sour on Web-based companies and, at the same time, our confidence in Surf N Chat's viability as a money-making venture had started to diminish.

Therefore, we abandoned our efforts and let Surf N Chat run unattended on the Web as a research experiment to see what happened. It started to slowly build up, by word of mouth, a user population. Unfortunately, we found that most of its use was for sharing adult content and (we think) for getting around company porn filters. The site was closed in 2002.

6 What Are You Doing For Us?

B Y THE LATE 1980s, it was clear that the Bell Labs culture had to change to address AT&T's business needs. AT&T was facing strong competition and not doing great financially. AT&T executives were questioning the value of Bell Labs Research. Circa 1989, according to Bill Brinkman, then head of physics research,[91] it was very clear that Bell Labs was in deep trouble. AT&T executives, "business suits," were asking Bell Labs executives

Hey, what in the hell are you doing for us?[92]

Unfortunately, despite such warning signs and Research chief Arno Penzias' realization in 1989 that Bell Labs should move away from academic research to helping AT&T, there was no wholesale strategic shift in the Bell Labs research strategy. By the early 1990s, many researchers were working on projects of potential interest to AT&T, but the number of such researchers constituted a small fraction of Research. Moreover, many of these projects were started without a buy-in from the AT&T business units. Sooner or later, it would be necessary to get the buy-ins from the AT&T business units and that was not going to be easy. Without buy-ins, the projects would go nowhere and would eventually be shelved.

Lucent executives were aware that Lucent's significant and profitable traditional voice telephone switching business was going to shrink in the coming years. Consequently, going forward Lucent would need a steady stream of new products, such as switches that could handle both traditional voice and IP data traffic and offer new kinds of services. The executives hoped that Bell Labs was going to be Lucent's key to success.

Lucent expected Bell Labs to develop new technologies and even products, enabling Lucent to charge ahead in the world of routers, IP telephony, IP PBXs, optical fibers, wireless, chips, and so forth. Lucent would have an edge over its competitors such as Cisco, Nortel, and Alcatel because of the mighty Bell Labs. Lucent competitors, not having an innovation engine like Bell Labs, would be forced to get new technologies by acquiring companies at inflated prices.

All this sounded wonderful for Lucent. However, there was the fly in the ointment that needed to be dealt with. Although Bell Labs was recognized as an innovation engine that had awed the world with technologies that had advanced science, Bell Labs now needed to do the same with technologies for innovative products.

Researchers believed that new technology was the be all and end all of research and that the business units should jump on their bandwagon and use their research to develop products. They were often surprised and dismayed when the business units would decline interest in their leading edge and patented technologies citing business reasons. Occasionally, some motivated researchers would go one step further and build innovative product quality [93] systems (as opposed to prototypes) that were enabled by their technologies. Even in case of systems that were potential products, provided a business case could be made, the business units would not typically climb on the bandwagon. They would instead pepper the researchers with questions such as how many customers do you have, what kind of support are you offering the customers, do you provide 24 × 7 customer support, what processes did you follow in system development, and what is the revenue potential.

These questions showed that the business units did not understand what Research could and could not do. Bell Labs was simply not geared to make and market products, find customers, offer full-fledged customer support, prepare business cases, etc. Business units were set up to do such things and therefore Bell Labs needed to collaborate with them.

Like the Indian fable of the "Blind Men and the Elephant," in which each of the six blind men feeling the elephant got a different understanding of what an elephant was, depending upon whom one talked to in the business units, one would get a different perspective about the role of Bell Labs Research. For example, many business unit managers wanted the researchers to develop products for them. Some business unit managers in switching wanted the research department in Naperville, IL to build demos with which to dazzle their customers. Other business unit managers wanted Research to do software development for them and looked upon the researchers as contract programmers.

Researchers believed, along with their managers, that their role was to innovate and excel in science. They expected the business units to develop products based on their research ideas. Business units, on the other hand, were simply not automatically interested in taking Bell Labs technology "nuggets" and making them into products. Business units, correctly, would undertake productization only if they could make a business case.

Lucent could get new technologies and products by acquiring companies, which it did often by paying exorbitant prices, or it could get them from Bell Labs. For the latter to happen, Bell Labs and the business units would have to work together.

However, changing the Bell Labs Research culture from that of freewheeling university-like research to one that met the business needs of Lucent in the fast paced Internet world was going to be a task of Herculean proportions. Before this could happen, many questions needed to be answered and understood. What would make the researchers and their managers change their behavior? What pressure would need to be applied to make the change happen? What incentives would be offered? Would the researchers leave rather than change their behavior? Did the researchers have the right skills to work on developing technologies and products for Lucent? Would the business units work with Bell Labs? How long would it take to change Bell Labs to an innovation engine that would help Lucent develop new products?

Some researchers, especially the system builders, upon being urged to build a product for the business units, said that they would instead prefer to leave Bell Labs. If they had to build products, they would rather do that in a startup or in Silicon Valley, because at least there they would get a large number of stock options and would have the potential of making big bucks.

Before the researchers could be convinced to change their focus, their managers had to be aligned with Lucent's business needs. Managers were jokingly said to have "bell-shaped" heads (in honor of the pre-divestiture AT&T logos) meaning that they were more likely to quickly align themselves with the interests of the company as compared to other employees. However, moving away from academic research to working with the business units was a fundamental change even for the managers. Such a change had started to happen in the last few years, but inculcating a culture in the managers of developing technologies and products for Lucent was going to take some more time.

In short, changing the Bell Labs research culture that had showered the world with so many inventions was not going to be easy. Unfortunately, the culture had to change quickly because otherwise Bell Labs would become irrelevant to Lucent. On the other hand, if the culture was changed too quickly and abruptly, many researchers might leave, including some marquis names, which would tarnish Bell Labs' reputation as the world's most famous research lab.

In addition to changing Bell Labs culture, there was also the very important issue of making the business units work with Bell Labs. Otherwise, Bell Labs researchers would not be working on the right topics and their technologies would not be of use to Lucent. Bell Labs had to become part of the business unit thinking. Changing the business unit culture would require the Bell Labs president or the vice president of research to work with the business unit leaders to make the change happen.

CHANGING THE RESEARCH CULTURE

Penzias started the process of change from academic research towards helping develop products and services for the AT&T business units by changing the mix of research and reorganizing Bell Labs to focus on AT&T's core competencies.[94] In 1990, Penzias closed research efforts in areas such as robotics and human interfaces that were not relevant to the post-1984 AT&T and expanded or started new efforts in areas that were potentially relevant to AT&T. Physical sciences research was cut from 80% of the budget to 50% while research in software and networking was increased to 50%. In addition, Penzias said that managers would be held accountable to ensure that their organizations worked on research topics relevant to AT&T. Some researchers felt that Penzias betrayed them by making these changes, which would move Bell Labs away from the sciences. They would also hold his successor Arun Netravali guilty of destroying long-term research because of his attempts to change the research model.

Netravali was aggressive in trying to change the research culture away from academic research towards helping the business units. He was at the helm of Research when Lucent was launched. Netravali had been appointed vice president of research on October 16, 1995, a few weeks after AT&T announced that it would split into three companies. Netravali was stepping into the shoes of his illustrious predecessor, the Nobel Laureate Arno Penzias, who had retired. Netravali knew that this was a critical time for Bell Labs:

> *I have been given a chance to lead a great organization at a time when the opportunity to make a difference has never been greater.* [95]

Netravali was also the president of Bell Labs during the turbulent times for Lucent, which saw the severe downsizing of Lucent and Bell Labs and his relinquishing the post of president. The call to become the president of Bell Labs had come in 1999 when his boss, Dan Stanzione, decided to retire. Upon becoming the president of Bell Labs, on October 26, 1999, Netravali said that the

> *... prospect of leading one of the world's premier R&D organizations is both electrifying and a bit daunting ... Bell Labs is a powerhouse of innovation. The collective brainpower here is awe-inspiring.* [96]

Technology was Netravali's forte and he thrived on it. For example, Netravali had the vision that by 2025, a century after the creation of Bell Labs, the earth would be covered by a communication skin consisting of

> *... millions of electronic measuring devices -- thermostats, pressure gauges, pollution detectors, cameras, microphones -- monitoring cities, roadways and the environment. All of these will transmit data directly into the network.* [97]

Such a communication skin would bode well for the Lucent telecommunications equipment business that, in 1999, was doing fabulously.

SHIFT AWAY FROM LONG-TERM RESEARCH

Netravali believed that the realities of the business world would shape research strategy and, consequently, he started steering Bell

Labs away from science and long-term research towards developing technologies and products for the company:

> *Since Arun Netravali took the helm a couple of years ago, there has been a shift away from the long-term [research] that created such milestones as the transistor.*[98]

Early in his career, Netravali realized that Bell Labs must help AT&T businesses, regardless of whether the work was classified as short-term or long-term research. This out-of-the-box thinking (for Bell Labs) was instrumental in propelling Netravali up the management ladder. The turning point in Netravali's career was the AT&T high definition TV (HDTV) project. Circa 1988, when he was the research vice president (old title was director) of center 1135, Netravali was instrumental in starting and leading AT&T's HDTV project. Netravali's expertise in video compression, a key technology required for HDTV, had made him an ideal candidate to lead the project.

In contrast to the analog broadcast TV standards in use today, e.g., NTSC in the USA and PAL in Europe and Asia, the digital HDTV standard would provide twice the current picture resolution and better interfaces to computers. The HDTV project would develop a proposal for an HDTV standard and a working prototype for submission to the Federal Communications Commission (FCC). If the FCC accepted AT&T's proposal as a standard, then AT&T would stand to reap significant revenues from the licensing and sales of HDTV components. AT&T would be competing with organizations such as Zenith, General Instruments, Philips Electronics, and MIT, who were also planning their own submissions to the FCC.

Although the HDTV project required the development of new techniques for digital image and video compression, the bulk of the work involved system development. Given Bell Labs culture, this project was looked down upon by many researchers and managers, some of whom were from Netravali's own organization. In addition, Netravali's boss did not view the HDTV project as a research project and was not in favor of Netravali working on it under the auspices of Research.

However, Netravali had started the HDTV project at the suggestion of and with the backing of the late Sol Buchsbaum, who was the executive vice president of Bell Labs, next only to the president. Netravali was thus able to deflect and ignore criticisms from his management, peers, and researchers.

Netravali built a team for the HDTV project and they worked long hours for many months to produce a prototype system, which they submitted to the FCC for testing. Subsequently, AT&T and Zenith combined their HDTV efforts to improve their chances of "winning." In 1993, the FCC recommended the merging of the various HDTV proposals. Subsequently, the "Grand Alliance" of the HDTV competitors proposed a standard for HDTV in 1995 and this was adopted by the FCC as the standard for the USA.

Netravali's gamble paid off. AT&T's HDTV project was a win for all of the parties in the Grand Alliance. This success would eventually take Netravali to become the president of Bell Labs. Had the HDTV project been deemed a failure, Netravali would have most likely been openly criticized as one interested in development instead of research.

Soon after taking over as the Bell Labs president, Netravali indicated his commitment to long-term research:

> *I think long-term research is very fragile ... The drive for speed, for faster new-product introductions, has sometimes led people not to invest enough in long-term research.* [99]

Through the 1990s, the public posture of senior Bell Labs leaders was that Bell Labs was committed to long-term research and advancing science. This posture was good for public relations and recruiting, and it kept the researchers happy with the senior leaders. However, privately, the senior leaders were telling the managers that they and the researchers should be developing technologies and products to help business units. Developing technologies and products was indeed the right policy since the parent company was in a competitive arena now and it needed innovations from Bell Labs to give it an edge over its competition.

However, because of the different public and private postures on Bell Labs research policy, the researchers received conflicting messages. With the senior leaders telling them to focus on science, they could not understand why some of their managers were telling them to focus on helping the business units. These contradictory stances confused the researchers and, as a result, most of them continued to operate as before.

Netravali was on the forefront of trying to change the research direction of Bell Labs. He targeted some researchers and managers to work on business unit related projects. He helped them start such projects. However, Netravali's actions were tactical. They did not constitute a strategic or wholesale shift in research direction, which undoubtedly would have caused many researchers to leave Bell Labs as happened at IBM for an apparently lesser change. In 1996, Paul Horn, upon becoming the director of IBM Research, changed its motto from

famous for its science and technology and vital to IBM

to

vital to IBM's future success

the implication being that science was not going to be as important as it used to be. As a result, according to Cherry Murray, now research senior vice president at Bell Labs, IBM lost 50% of its physics researchers.[100]

By 2001, because of its difficulties, Lucent needed Bell Labs even more than before. It was imperative that researchers be given an unambiguous message that their job now was to help the Lucent business. Addressing a meeting of about 75 researchers in the spring of

2001, Jeff Jaffe,[101] then research senior vice president, said words to the following effect:

> *For us to do long-term research in the long term we have to do short-term research in the short term.*

Jaffe told the researchers that they had to do short-term research and work on topics of interest to the Lucent business, at least while Lucent was going through tough times. Research topics had to be aligned with the Lucent business. Some researchers could work on other topics provided the research was expected to produce significant results. For example, if a researcher was not aligned with Lucent business needs, then the researcher's work had better be of Turing Award [102] quality. Surprisingly, Jaffe's remarks did not even raise a whimper of protest. In days gone by, Jaffe's remarks would have caused uproar. Researchers knew that the survival of both Lucent and Bell Labs was at stake. For many researchers, leaving Bell Labs was no longer an option since the economy was in a downturn and many companies were downsizing. The dot-coms had gone bust, and the telecommunications and software industries were in a deep slump.

Netravali was always deeply involved in a few key research projects. He would push researchers and managers to start new projects, often suggesting topics. Netravali was focused on ensuring that research projects had business value. He would ask generic probing questions, such as what was innovative or novel about a project, what was its business value, and who in the business units cared about the project, and he would then demand clear and satisfactory answers from the researchers and their managers. Many of them felt uncomfortable answering these questions, since they were not used to justifying their research, particularly with respect to business value. Even I was uncomfortable in the beginning. In my first six or seven years at Bell Labs, nobody had ever asked me such questions. By the time I became a manager, asking these questions of the researchers was ingrained in me.

Netravali would initiate meetings with the business units, write follow-up letters, and so forth. He started doing this early in his career and continued to do so even as the president of Bell Labs. Few of Netravali's managers could come close to matching him in his zeal to help the researchers connect with the business units.

Netravali personally mentored several key research projects. He would work directly with the researchers and their managers, often bypassing two or three levels of management. Netravali mentored the projects because of their potential value to the company and because he wanted to move them at Internet speed.

In 1996, to promote cooperation with the business units, Netravali, then vice president of research, and Bell Labs president Dan Stanzione devised the concept of "breakthrough" projects. If Netravali and a business unit president agreed that a research project was both innovative and important for the business unit, then it would be designated as a breakthrough project and would be jointly sponsored by them. Research would contribute researchers and additional resources as appropriate, while the business unit would provide development and business resources. Since both Research and a business unit would be backing a breakthrough project, it was hoped that such projects would have a higher chance of business success.

Netravali appreciated, encouraged, and rewarded researchers who worked on projects to help the company. However, many researchers and managers hated Netravali for attempting to turn Bell Labs away from long-term research, thus destroying their comfortable world of academic-style research.

UPPING RESEARCHER QUALITY

Netravali placed a high emphasis on researcher quality. He wanted to make sure that the managers hired the best candidates. Consequently, he would scrutinize each hiring request, often rejecting those that he thought were subpar candidates such as those who were not from the top universities or who came with less than stellar recommendations. Netravali often expressed a desire to weed out the bottom five or ten percent of the researchers every year, based on the annual performance review. Here he did not have much cooperation because managers did not like this unpleasant task and because the Bell Labs research culture did not include firing people.[103] Nevertheless, Netravali's persistence eventually started paying off with managers moving to weed out the underperforming researchers.

Netravali was especially critical of the research managers in the Bell Labs locations at Holmdel (NJ) and Indian Hill (Naperville, IL) because of his perception that the average researcher quality at these locations was lower than that at Murray Hill. There were two important reasons for this. First, it was harder to get the top researchers to work at Holmdel or Indian Hill because they wanted to be at the prestigious Murray Hill location, where most of Research and the famous researchers were located. Second, the Holmdel and Indian Hill research organizations were co-located with the business units. Because of personal connections, several business unit employees had been able to transfer to Research. Many of them would not have met the criteria used for hiring people from outside the company.

As soon as the Holmdel and Indian Hill organizations came under his purview, Netravali insisted that the weak researchers be weeded out. The Holmdel and Indian Hill managers were not enthusiastic about doing do this. As a result, Netravali applied direct and indirect pressure to make the managers cooperate. Not surprisingly, Netravali was not popular with the researchers and managers at Holmdel and Indian Hill.

WORK LIKE A PROFESSOR, GET PAID LIKE A PROFESSOR

Except for teaching, a small part of a professor's job at most topnotch universities, and writing grant proposals, the work of an "academic-style" researcher at Bell Labs was similar to that of a professor. However, the researchers at Bell Labs were paid much more than professors.

Researchers doing academic-style research were by and large more famous than the researchers building systems and working with the business units. They were more famous because they had more opportunities to publish papers, be on program committees, go to conferences, etc. Unfortunately, researchers working on company projects generally did not get such external visibility.

In determining researcher compensation, Bell Labs did not make a distinction between the academic-style researchers and those working on projects to help the business units. In fact, the external visibility of the researchers publishing papers actually helped them get better raises. As a result, there was no incentive for the academic-style researchers to work on business unit related projects.

However, there was really no reason, besides working at Bell Labs, why the academic-style researchers were being paid more than university professors. On the other hand, researchers helping the company business, for example, by developing new products, needed to be paid more to bring their compensation in line with what they might get at a startup. This was in the heyday of the dot-coms and by doing this Bell Labs might also be able to stem the flow of the best researchers to startups.

In such a scenario, salaries of the academic-style researchers would be less than those of the other researchers. This differential would create an incentive for researchers to work on projects that would help the company. To create the differential, the compensation of the academic-style researchers would have to be brought down to match those of university professors. However, who was going to "bell the cat?" This was not going to be an easy task since a two-tier salary would create a big furor in Bell Labs.

Late in 1999, Shamim Naqvi was appointed research vice president of center 1138 (the research organization in which I was a director). Talking about compensation to the members of 1138, Naqvi said that from now on compensation would reflect value to Lucent. Researchers focusing on science and publishing papers would get competitive academic salaries and researchers building systems useful for the Lucent business would get compensation similar to what they might get in startups.

This was the first time a senior manager had openly announced moving towards a two-tier salary structure for the researchers. After hearing Naqvi, many researchers became visibly agitated because they felt that the academic-style researchers would become second-class citizens. Within a few days, one of my academic-style researchers resigned to go to AT&T Labs and then another left shortly thereafter to go to Korea.

A few days later, to show that he was serious, Naqvi promised to give the researchers working on two projects with the business units extraordinary bonuses. Naqvi promised the project team members up to 100% of their salary as bonus for completing the projects to meet an aggressive schedule. The bonus would be paid in installments, as the teams met their milestones.

Naqvi left Bell Labs a short time later. Nevertheless, the researchers who had been promised bonuses did get them.

The two-tier compensation policy was not put into place. By 2001, the academic-style research and the two-tier compensation issue had become largely moot. With jobs scarce, technology and telecommunications industries downsizing, and the dot-com fever gone, most researchers were happy to have jobs and were quite willing to work on business unit related projects.

RESEARCH TRIES THE STARTUP MODEL

Business Week, in an article titled "Masters of Innovation," referred to the Bell Labs research model, circa 1950, as the traditional model of innovation in which

> ... *genius-scientists locked away in a lab devising new materials, components, and software, which their employer would then patent*

and eventually use to improve products or create new ones. ... the more patents a company churned out, the greater its bragging rights on innovation.[104]

This research model was used in Bell Labs through the 1980s. Starting in the 1990s, Bell Labs leadership started to change the research model. Responding to the *Business Week* article, Netravali wrote a letter saying that Bell Labs

> *... has come a long way since then in its approach to nurturing innovation. There used to be a number of layers between Bell Labs and the customer. ... Now development teams work directly with customers during the design phase. We also often now use customer networks to beta-test new products ... Bell Labs innovations are making the journey from lab to market faster than ever ...*[105]

Unfortunately, only a handful of research projects were working directly with the customers. Such projects were exceptions rather than the rule. In addition to the breakthrough project concept for collaborating with the business units, Netravali had simultaneously tried another approach. Some research projects would operate like startups. Instead of just developing the technology or a prototype and hoping that the business units would make it into a product, Bell Labs would develop the product and do so while working directly with potential customers. Only after customers were in place would Bell Labs hand over the research project to a business unit.

The researchers would work with customers from the very beginning during the requirements process, the design and implementation process, and field trials. And they would possibly take their system to initial commercial deployment. The research teams would operate at Internet speed bypassing the "slower" business units with their numerous processes such as feasibility studies, market testing, focus groups, documentation, and code hardening. Customer field trials would be a faster substitute for such business unit processes. The research teams would work with the business units, especially their marketing folks, but the projects would be under the auspices of Bell Labs.

The following three projects, PacketStar, PathStar, and Softswitch are examples of the startup model. Each project represented a disruptive technology for Lucent. Although there was cooperation from the business units, the projects were controlled by Bell Labs. Each of the projects brought products to the market rapidly, in a year or less.

PacketStar

The PacketStar™ was a next generation intelligent super fast router, which in technical terms would be specified as a layer 3/layer 4 wide area network 128 Gigabit router. It was developed in twelve months by a team led by Vijay Kumar. The PacketStar team had been started by Kumar with the backing of Netravali. To ensure that PacketStar would have proper business support, the PacketStar team was moved from Research to the Data Networking Systems (DNS) business unit,[106] a natural home for the PacketStar. Kumar's team would work on the core router while other teams in DNS would work on related software and the surround required for making PacketStar into a product.

PacketStar seemed to be on the road to success and was going to put Lucent in the forefront of IP product space. According to Bill O'Shea, then president of Lucent's DNS business unit,[107] PacketStar was

> ... *truly a revolutionary switch ... For the first time, one system combines the superior bandwidth and speed to help break through Internet data bottlenecks with intelligent traffic management capabilities to guarantee multiple tiers of service.*[108]

Carly Fiorina, then president of Lucent's Global Service Provider Business,[109] said in May 1998 that PacketStar's

> ... *capabilities are critical to delivering large-scale telephony and fax services over IP networks, data and video over IP and virtual private*

networks. For the next-generation applications that customers de-mand, the carrier-class PacketStar IP Switch is essential.[110]

The research team had operated at Internet speed to develop PacketStar. More work needed to be done before PacketStar could be marketed as a product. Unfortunately, PacketStar was not destined to have a long life. In November 1999, Lucent acquired Nexabit Networks, a developer of high-performance IP switching equipment.[111] Nexabit's product line overlapped with PacketStar and this led to the cancellation of PacketStar.

PATHSTAR

The PathStar™ Access Server, a central office switch-router that seamlessly connected circuit and packet networks, was designed to enable network operators to offer low-cost, reliable voice and data services over Internet (IP) networks. PathStar converted everything to IP packets and sent them over an IP network (private IP networks or the Internet). It offered the combined functionality of a Class 5 telephone switch,[112] a digital loop carrier (bundles multiple telephone lines), VoIP gateway for Internet telephony, a voice announcement server, and an edge router.

PathStar was revolutionary because, according to Bill O'Shea, it offered an

... integrated routing IP platform that services most types of traffic – from POTS and ISDN to xDSL – and delivers on the features customers demand, from billing to directory services.[113]

Phil Winterbottom, who led the design and development of PathStar, was the heart and soul of PathStar. It was developed in record time, the research team taking about ten months to develop PathStar and bring it to the market.

PathStar was a very attractive system for the CLECs,[114] the small new local telephone companies that were springing up, which could use one system to offer a variety of services. PathStar supported up to 10K subscriber lines and was thus more suitable for the CLECs as opposed to the big ILECs,[115] the established local telephone companies such as Verizon. PathStar was an inexpensive switch, costing between $250K and $500K depending upon the discount, though a fully loaded system could cost much more.

PathStar sales to the CLECs did well, about $100 million a quarter. However, they started faltering when many of the CLECs started getting into financial trouble, which eventually led to their collapse circa 2000. Lucent and other equipment makers had to write off hundreds of millions of dollars in loans when the CLECs went bankrupt. This led to the cancellation of PathStar in early 2001 in favor of another switch, the higher end 7R/E.

SOFTSWITCH

The Lucent Softswitch, originally known as the Saras Softswitch, was an innovative switch that seamlessly connected the traditional circuit switched and IP networks. It aimed to combine the reliability and features that customers expected from the public telephone network with the cost effectiveness and flexibility of IP technology. The Lucent Softswitch also provided open APIs to allow third parties to easily build new applications, which was not possible with traditional switches.

Softswitches represent the next generation alternative to traditional switches, which can be expensive and difficult to extend with new features. With a softswitch, IP service providers can offer telephone service (using the Internet telephony technology VoIP), and, using softswitch APIs, they can offer new services. Softswitches are considered the holy grail of voice switching since they can bridge IP networks with traditional voice networks,[116] thus allowing telecommunications companies to use IP networks without jeopardizing their investment in the traditional voice networks.

The primary idea underlying a softswitch is the separation of call control and signaling from media and access gateways. A traditional switch must be purchased from one vendor while a softswitch and related hardware can be purchased from multiple vendors. A softswitch is cheaper to deploy than a traditional switch because it can run on standard off-the-shelf servers and because media and access gateways are commodity items.

The Lucent Softswitch research team was led by its creators, Murali Aravamudan and Shamim Naqvi. It was primarily the brainchild of Aravamudan. Naqvi was a very eloquent and forceful spokesperson for the Lucent Softswitch. They worked well together as a team and brought the Lucent Softswitch to market in less than a year. The Lucent Softswitch team worked with potential customers such as Level 3, a communications provider, to ensure that it would meet customer needs.

The first sale of the Lucent Softswitch was to Level 3, which planned to use the Lucent Softswitch as the foundation for its next generation broadband networks. Speaking about the Lucent Softswitch, James Crowe, president and CEO of Level 3, said

> *The Lucent Softswitch will deliver high-quality voice over the Level 3 IP network. It will let us provide the kind of services our customers want from circuit-switched networks, and it lays a foundation on which our entrepreneurial partners will build a new generation of innovative broadband services.* [117]

The sale of the Lucent Softswitch was part of a $250 million dollar strategic agreement in which Level 3 was going to buy Lucent systems including the Lucent Softswitch. According to Lucent and Level 3, this agreement had the potential of growing to $1 billion over five years.

Netravali, then vice president of research, was quick to publicize this deal within Bell Labs as a big success. A research team had worked directly with customers to develop a new type of switch and represented a clear example of how Bell Labs was turning around to help Lucent. The Lucent Softswitch team was held up as a paragon to be emulated by other researchers.

Although the Lucent Softswitch was a small part of the Level 3 deal, the research team claimed that without the Lucent Softswitch the Level 3 deal would not have occurred. Aravamudan and Naqvi were thus, they claimed, instrumental in helping increase Lucent revenues.

The Lucent Softswitch got very good press outside Lucent, for example, in *Business 2.0*:

> ... *the most impressive product to come out of Lucent in the past year – and one that indicates Lucent may be able to make the transition to a Valley data-centric company – is softswitch.*[118]

Because of this research success, Netravali promoted Aravamudan and Naqvi to vice presidents. Aravamudan became the CTO in the Communication Software business unit reporting to Bob Madonna, the former CEO of Excel Switching Corporation that had been recently acquired by Lucent. Naqvi was promoted in research to manage the Communications Software Research Center (organization 1138).

Soon after the Lucent Softswitch deal, Netravali was himself promoted to become the president of Bell Labs (upon the retirement of Dan Stanzione). Many researchers believe that the Level 3 deal was a key factor in Netravali's elevation to Bell Labs president, since it projected him as a leader who was taking Bell Labs technologies at Internet speed to the marketplace.

The Excel Switching Corporation, a leading developer of programmable switches, was acquired by Lucent in August 1999 for about $1.7 billion in stock.[119] The acquisition was completed on November 3, 1999. Excel brought with them their own softswitch. Because of the Excel acquisition, Lucent now had two softswitches. From a business perspective, Lucent could market only one softswitch as a product. It was decided that the Lucent Softswitch would be merged with the Excel Softswitch by putting some of its features into the Excel Softswitch. The Excel Softswitch, billed as the second-generation Lucent Softswitch, was unveiled by Lucent in January 2001.[120]

RETROSPECTIVE

These three projects represented a radical change in the traditional Bell Labs research model, away from the traditional way of doing research. The researchers were involved in all phases of product building, such as conception, meeting customers, requirements, development, field trials, and initial deployment. The leaders of these projects, Kumar, Winterbottom, Aravamudan, and Naqvi were all very motivated to build products for Lucent. They pushed their teams hard because they wanted to be the first to market with their systems.

Unfortunately, these projects were negatively affected by Lucent's acquisitions and the downturn in the telecommunications market. These events were beyond the control of the researchers. Although the projects were canceled, the researchers had shown that they could change with the times. Bell Labs was adapting its research model to meet the needs of its parent company.

COLLABORATING FOR SUCCESS

Not every research project needed to or had to operate as a startup to successfully produce a technology or develop a system for the Lucent business units. Depending upon the technology and business needs, a simple collaboration between Bell Labs and the business units could also be very effective. For example, circa 1998, I persuaded one of my researchers to join me in investigating LDAP directories (specialized lightweight databases). Cisco and others were starting to promote the use of LDAP directories in networks to facilitate tasks such as equipment configuration, network management, and interoperability of network applications. Directory-enabled networks (DEN), networks that use directories, were going to be the basis of intelligent networks.

My next step was to find some projects in the business units where directories could be useful and where the business units would be receptive to working with Research. Eventually, we found an organization in the Business Communications Systems (BCS) business unit[121] that needed to develop a common directory that their systems could share and use together.

After understanding BCS needs and taking a proactive stance, we made a preliminary proposal for a directory integration project that would unify data from multiple directories. The BCS manager, with whom we had the initial discussions, agreed that our proposal would help BCS in simplifying PBX administration. However, over the next many months, not much happened. There were more meetings, but BCS was noncommittal about starting a joint project with Research.

In early 1999, Ravi Sethi,[122] who at the time was a senior Research executive, heard about our proposal to collaborate with BCS. He stepped in to talk to his counterparts in BCS and helped establish a joint project with BCS, which would contribute staff to work with the Research team. I was particularly pleased to learn that our BCS contact would be Hector Urroz, a manager whom I had known for some years and who had always been enthusiastic about working with Research.

Specifically, the Research/BCS collaboration involved integrating the directories of two systems, a PBX and a voice messaging system. Although these systems worked with each other, they needed to be better integrated. Each system had its own directory of employee information. When a new employee came on board, or one left, or the information about an employee changed, each directory had to be updated independently. Not only was this a bit absurd,[123] it also led to inconsistencies between the two directories because sometimes inadvertently different data was entered in the two directories or data was entered only in one directory.

The solution involved building a new interface that would automatically update directories in both systems. This was simple. The challenge was in ensuring that the solution would support existing interfaces so that current processes and applications would continue to work as before. The solution had to ensure that any change made to one directory would be automatically reflected in the other directory. Finally, the solution would have to eventually support the integration of more than two directories.

At this time, I was able to also interest two more researchers in the directory integration project. The three researchers worked wonderfully with the BCS team. The combined Research/BCS team, worked hard and enthusiastically for several months to produce a system, called MetaComm, for integrating data from various telecommunications devices. MetaComm used Bell Labs technologies such as those for updating integrated views of multiple (independent) databases and LDAP alerting functionality (I was one of the creators of the latter).

MetaComm was a success from both business and research perspectives for several reasons. Most importantly, MetaComm was used in a real product. In addition, two research papers describing the innovation used in MetaComm were published and one patent was filed.

The collaboration itself was successful for several reasons. Research did not start the project on its own and then throw it over the wall to BCS. After identifying a potential area where Research could help BCS, the Research team worked to establish the collaboration with BCS before starting the project.

Paralleling this was the fact that BCS needed technology to integrate directories of its PBX and voice messaging systems that would preserve existing application interfaces. BCS did not have the expertise to invent the technology on its own. As a result, they were happy to collaborate with Research, invest resources in the project, and teach the researchers the specifics of the PBX and the voice messaging systems.

Despite some initial hesitation, the researchers were flexible and willing to work on a project that was of interest to the business unit. They did not have any hang-ups about whether or not this project would lead to research publications. Their focus was on helping BCS. The researchers, like the business unit team members, were hands-on software experts who were not afraid of getting their hands "dirty" by building a real system. The researchers were especially motivated by the fact that BCS would use their system in a real product.

An important reason for the success was that both Bell Labs and the business unit collaborated voluntarily. Sethi's help as a facilitator was important. In addition, I had a good relationship with BCS, based on failed past collaboration attempts. Moreover, Hector Urroz, the BCS point of contact, was enthusiastic about cooperating with Research.

7 Bell Labs Goes West

O N JULY 1, 1998, Lucent announced that Bell Labs was establishing a new research organization in the heart of Silicon Valley, in Palo Alto, CA. This new arm of Bell Labs was dubbed Bell Labs Research Silicon Valley (BLRSV). It was to focus on data networking and, according to Dan Stanzione, president of Bell Labs and chief operating officer of Lucent, the new Bell Labs outpost in California would

> ... *blend Bell Labs innovation and technical excellence with the unique entrepreneurial culture of Silicon Valley.* [124]

Silicon Valley was the hotbed of action in new technology. New startups were sprouting like wildflowers and taking on the established giants of telecommunications technology such as Lucent and Nortel.

Lucent wanted to operate like a Silicon Valley company, as an aggressive start up and not as a stodgy offspring of Ma Bell. Lucent wanted to be part and parcel of the Silicon Valley culture so that it could move aggressively and quickly to introduce new products in the telecommunications market. Lucent hoped to take on the likes of Cisco and make sure that Microsoft was not tempted to enter Lucent's telecommunications world.

Bell Labs Research [125] management wanted researchers to create new technologies and products for Lucent at Internet speed, the pace at which new products were being created by Silicon Valley startups. They wanted BLRSV to operate like a Silicon Valley incubator that nurtured and launched companies that quickly converted research ideas into products. Meanwhile, Lucent was making its presence felt in Silicon Valley by buying Silicon Valley companies such as Octel Communications and Livingston Enterprises, and setting up a venture fund, Lucent Venture Partners, in Silicon Valley.

Bell Labs Research in New Jersey could definitely benefit from the startup culture of Silicon Valley. There was a wealth of talent, technical combined with entrepreneurial, in Silicon Valley that could not be attracted to move to New Jersey. Moreover, Bell Labs needed to plant its flag in Silicon Valley so that it could become part of the Silicon Valley culture and work with the companies Lucent was acquiring in California.

BLRSV was to work with Lucent business units in Silicon Valley and to have a close relationship with Lucent Venture Partners. To ensure the latter, BLRSV would be co-located with Lucent Venture Partners.

NOT CLONING MURRAY HILL

From the outset, BLRSV was planned primarily as a non-traditional research organization, one that embodied the Silicon Valley startup culture. In particular, BLSRV would be very different from the more established and traditional research organization, Bell Labs Research, back at Lucent's headquarters in Murray Hill, NJ.

BLRSV would be product focused, with the explicit goal of producing new and innovative technologies and products for the Lucent business units. Unlike their counterparts in Murray Hill, BLRSV researchers would get "phantom" or virtual stock (or some kind of profit sharing) in each project that was trying to build a product that could become a Lucent offering. Projects building products for Lucent businesses would thus be treated as "internal ventures."

Lucent business units would be the preferred acquirers of these internal ventures. They would have the right of first refusal. The phantom stock associated with an internal venture would convert to cash if and when the internal venture was bought by a Lucent business unit. The valuation of the internal venture would be close to market valuation. If no Lucent business unit was interested in the internal venture, then BLRSV would attempt to spin off the venture as a separate company.

Offering equity to the researchers in research projects was a radically new idea for Bell Labs and for the research community in general! By doing so, BLRSV hoped to generate in its researchers the same sort of enthusiasm and passion that permeated the startups. In effect, BLRSV would be incubating companies for the Lucent business units. BLRSV would help researchers in creating these internal ventures by providing them with the necessary financial, technical, and business resources. For example, BLRSV would provide staff for specifying product requirements and doing market analysis.

Providing researchers with business resources was a new direction for Bell Labs. Researchers at Murray Hill did not traditionally have direct access to staff with business skills. Bell Labs Research did not have staff on its payroll to help researchers with business functions such as product specification or market trials. However, research chief Arun Netravali was trying to change the culture in Research by ensuring that key research projects did not operate in a business vacuum. For example, when researchers established collaborations with business units, Netravali would expect, even require, the business units to provide resources to help, besides other things, in product definition, making a business case, and market analysis. Moreover, in 2001, Netravali also approved the hiring of a few business analysts to help ensure that projects in Research were addressing real business needs.

For the first time Bell Labs researchers, especially those located at BLRSV, with entrepreneurial ambitions would have a path to quick riches from within the confines of Bell Labs. The motivation to convert phantom stock into cash would hopefully galvanize the researchers into exploring and developing products of value to Lucent business units. Lucent hoped that the researchers would put in the long hours necessary for making the internal ventures successful. Because of the financial incentive, Lucent also hoped that the researchers would think "out of the box" to come up with innovative technologies and products that would help Lucent leapfrog past the competition.

Lucent was not being altruistic in offering the researchers phantom stock and encouraging them to create these internal ventures. Lucent hoped that this entrepreneurial arrangement would allow it to acquire new technologies and products at a significantly lower cost than what it would have to pay to buy companies in the Silicon Valley landscape (or elsewhere).

To succeed, BLRSV had to have both internal connections within Lucent and external connections with Silicon Valley. The internal connections were necessary for BLRSV to connect with the rest of Bell Labs back east and with the Lucent business units. The external connections were necessary so that BLRSV could hire the best in Silicon Valley for the internal ventures. With internal and external connections in mind, on July 13, 1998, Bill Coughran and Ron Schmidt were both appointed as research senior vice presidents to co-head BLRSV. Coughran brought (Lucent) internal connections while Schmidt brought external connections to the table.

Bill Coughran came with an extensive background in research and much experience in research management, acquired during the nearly twenty years he had spent at Murray Hill. Before his promotion to co-head BLRSV, Coughran had been the head of the famed UNIX center in Murray Hill where the UNIX system had been developed. Because of his long years in Research, Coughran had strong ties with and connections in Bell Labs. Coughran also had many connections in the business units. In addition to his Research job, Coughran had also been the chief technical officer (CTO) in Lucent's Communications Software Group business unit.

In contrast, Schmidt came with an entrepreneurial background. Schmidt was very well known and respected in Silicon Valley. He had been a co-founder and CTO of SynOptics, which had merged with Wellfleet Communications to form Bay Networks. He was also an advisor to and a board member of many Silicon Valley startups. Schmidt brought with him strong connections to the Silicon Valley entrepreneurial world, a valuable asset that would help BLRSV in hiring the best people to staff the internal ventures.

BUILDING BLRSV

Bell Labs Research typically hired freshly minted PhDs with a focus on academic research and publications.[126] With the changing needs of the corporate parent of Bell Labs following the 1984 breakup, hiring, for example, had grown more selective in terms of acceptable disciplines. In the past few years, at the behest of research chief Netravali, emphasis was being placed on hiring PhDs with system building experience. But hiring was still primarily based on academic excellence with the hope that, once on board, the researchers would work on Lucent business problems.

BLRSV would focus on hiring researchers with innovative ideas for new products, skilled in building systems, and with entrepreneurial ambitions. Having a PhD would not be a requirement and traditional academic excellence and publications in top journals, etc. were not going to be issues. However, hiring such researchers was going to be difficult in Silicon Valley's very competitive environment. To ramp up, Bell Labs would also hire some traditional researchers, which would be relatively easy given Bell Labs' reputation. And BLRSV would hire a few people with business skills.

BLRSV would use the phantom stock concept to attract the best "entrepreneurial" researchers. This would make BLRSV's compensation plan competitive with the compensation plans followed by companies in Silicon Valley, plans that offered employees the opportunity of getting rich quickly. Bell Labs Research, like the rest of Lucent, followed the traditional employee compensation plan, similar to the ones followed by many of the big established companies. The key components of the compensation plan consisted of items such as a salary with an annual increase based on performance and an annual bonus. The latter had two components, a merit based bonus and a bonus that was a fixed percentage of salary, the percentage depending upon Lucent's performance. Key employees would also get a small number of Lucent stock options. (The number of options granted has been increasing.) And all employees would automatically participate in a pension plan that allowed a qualified employee to retire with an annual pension.

A key component missing in the above compensation plan was the potential of a big financial upside that was possible in the compensation plans offered by startups. Even if a researcher worked hard, put in long hours, and helped develop a product that was a big commercial success, the researcher's upside was limited. In contrast, for example, Wall Street[127] rewarded its employees with enormous bonuses and startups gave their employees large numbers of stock options.

Until a few years ago, Bell Labs researchers were not given stock options or large bonuses. These perquisites were reserved for senior management. Bell Labs researchers started getting options after the birth of Lucent, but until circa 2000 the number of options given to them was small.

A handful of research teams were working aggressively to develop new products for Lucent. These teams needed to operate like startups to develop products quickly. Netravali wanted to reward performance and encouraged the teams to move ahead rapidly. Team members were promised and reportedly received large stock option grants. These grants were also instrumental in retaining key people at a time when many were leaving for the lode of gold at startups. Unfortunately, with Lucent's stock price going down, most options have now become worthless, at least for the near future.

Bell Labs leadership hoped that researchers with an entrepreneurial bent would find BLRSV an attractive environment in which to build new products. Giving them a stake in the creation of successful products would ensure a steady stream of new products for Lucent's business. Besides offering an upside similar to that offered by startups, working at BLRSV would have many additional advantages. For example, the founders would not have to look for seed money since BLRSV would pay them and others (good) salaries, provide space, equipment, and other resources necessary for starting a new venture. The founders and other team members would get equity as in a normal startup. There would be no downside in case of failure since the founders and team members would continue to be employed by BLRSV while they explored the formation of another venture. Finally, the Lucent business units, assuming an innovative product was built, would be potential acquirers of the internal venture. The internal ventures thus had tailor-made acquirers waiting in the wings.

However, joining BLRSV to create a Lucent internal venture did have some disadvantages in comparison to getting venture money to create an independent startup. Potential founders would have less independence than in a startup because they would have to operate under Lucent's constraints and culture. Founders would also get less equity than if they started the company independently. Lucent would keep a significant percentage of the equity because it would be providing the founders with money from day one and would be limiting their downside. In addition, if a Lucent business unit was not interested in an internal venture, it could put roadblocks in spinning off the internal venture as a separate company.

Overall, however, the internal venture concept seemed to be a good deal for a researcher with entrepreneurial skills. BLRSV offered an attractive environment for starting a new venture for technical people who had bright ideas, but were somewhat risk averse.

Staffing BLRSV was challenging from the start. BLRSV, when fully ramped up, was expected to have about 125 employees. Hiring in Silicon Valley had become very competitive because these were amazing boom times for entrepreneurs. Armed with a PowerPoint presentation, one could easily be on the road to starting a venture. Anyone with a half decent idea could easily find venture money through angel investors and venture capitalists. The lure of quick riches was causing problems in the research and development arms of many established companies, particularly in Silicon Valley. Many researchers and technical personnel, especially those with the expertise and skills necessary for developing products, started or joined new ventures. Even Bell Labs back in the east had lost researchers because of the startup fever.

Hiring top-notch technical people with entrepreneurial ambitions was going to be difficult. Many of the entrepreneurs were risk takers who wanted much more control of their own destiny than BLRSV and Lucent could offer them. Nor were they likely to live with the restricted upside offered by BLRSV or any other company. Joining BLRSV to start an internal venture with the primary exit strategy being to convince a Lucent business unit to buy the venture was not a compelling scenario for the best entrepreneurs.

However, with Schmidt, BLRSV was now connected to Silicon Valley. He was the key to BLRSV's recruiting in Silicon Valley. Most of the BLRSV employees were either recruited by Schmidt or came as a result of Schmidt's connections to Silicon Valley. At the peak of its short life, BLRSV had within a year hired about 30 employees, including some traditional researchers and several contract programmers.

The BLRSV research model soon started encountering some challenges. There were delays in getting approvals from Bell Labs back in Murray Hill, NJ to hire staff with business skills (as opposed to technical skills). Bell Labs Research had traditionally hired technical people (besides support staff) and there was resistance to hiring people with business skills as mainstream employees. From the BLRSV perspective, staff with business skills was an important ingredient of the internal ventures.

Nevertheless, moving forward, BLRSV started several projects with the intention of developing products for the Lucent business units. Difficulties arose with the concept of selling the internal ventures to the business units. The business units felt that they should not have to pay to acquire the internal ventures since they were the ones funding Bell Labs and thus the internal ventures.

BLRSV management started getting frustrated because their internal venture model of delivering products to Lucent was not working. Since the business units were not interested in or agreeable to acquiring the internal ventures, BLRSV moved to spin off internal ventures as independent companies. BLSRV would work with the Lucent New Ventures Group (NVG), located back in Murray Hill, to spin off the companies. NVG had been set up to help create new companies around Bell Labs technologies that were not of interest to the business units. Working with NVG, BLRSV did spin off two companies in 2000: NetCalibrate, which built tools for measuring the quality of service on a website, and nSolutions, which built tools for automating the deployment of multi-vendor IP infrastructure.

However, a potential BLRSV optical internal venture that was in the process of being spun out ran into problems with a business unit. The business unit, which had competing technology, was not prepared to buy the venture nor was it agreeable to letting it be spun off as a separate company.

Apparently frustrated, Schmidt left BLRSV in early 2000. With him left BLRSV's connections to the Silicon Valley entrepreneurial world. Many of the people that Schmidt had recruited started following him in leaving BLRSV. Senior leaders of Bell Labs Research, back in New Jersey, began to feel that the BLRSV model of internal ventures was not workable in the context of Lucent. They had not been able to work out the details of a compensation plan that would be acceptable to them and to the BLRSV researchers. The implementation of the phantom stock idea, which had not yet been put into place, was abandoned in mid 2000. Senior Bell Labs leadership decided that BLRSV would instead follow the traditional compensation plan used by the rest of Bell Labs Research.

Coughran was apparently not happy with the scenario that had evolved at BLRSV since he wanted to create new companies. Operating BLRSV as a smaller clone of Bell Labs Research in New Jersey did not appeal to him. Coughran followed Schmidt and left BLRSV in the fall of 2000, leaving BLRSV leaderless.

A BOLD EXPERIMENT CUT SHORT

BLRSV's internal venture model was a bold experiment by Bell Labs leadership to try to galvanize researchers into building, at Internet speed, innovative products for the business units. The internal venture model was new to the business units and to Bell Labs. This raised some issues since now the business units would have to deal differently with Bell Labs. They were "used" to getting stuff "free" from Bell Labs. Now they would have to "buy" stuff from Bell Labs, from an organization they were funding almost completely. The business units needed to be convinced that in the long run it would be cheaper to acquire the internal ventures than acquiring companies in the open market.

Making the internal venture model work in Lucent would require some time. However, it was not "cost effective" for the best BLRSV employees to mark time while the internal venture model issues were resolved. In the extremely competitive Silicon Valley, it was easier for them to go elsewhere or start their own company since all they needed was a PowerPoint presentation.

BLRSV morale headed lower with successive departures of Schmidt and Coughran, and some key employees. Towards the end of 2000, several months after Coughran's leaving, Bill Brinkman, vice president of research, asked me to take charge of BLRSV. I declined because personal constraints precluded my moving to California. This was a difficult decision for me since I knew that I was turning down an important promotion, an opportunity that might not come my way again. However, I was persuaded by Bell Labs president Netravali to head BLRSV with the understanding that I would be bicoastal, commuting between California and New Jersey since I would also be responsible for one department in Murray Hill.

Visiting BLRSV for the first time at the end of 2000, I found it in disarray. Many staff members had left and only about twenty remained. Recruitment in 2000 had been very difficult. Top-notch system builders and entrepreneurs were reluctant to join BLRSV while BLRSV did not want to hire theoreticians.

My first step in understanding the state of BLRSV was to have a one-hour meeting with each staff member. There were many interesting projects going on at BLRSV. Researchers were working on several security related topics, fiber to the home, wireless, IP networks, etc. Unfortunately, these projects were all operating in a vacuum with respect to the business units and the rest of Bell Labs.

By the end of the second full day of meetings, it was clear that BLRSV needed, in addition to leadership, connections with the Lucent business units and with Bell Labs in the east. A strong effort had to be made to recruit researchers interested in building systems. Despite these problems, many of the BLRSV researchers wanted BLRSV to succeed. I had to rely on some of them to help me nurse back BLRSV to health. I was going to promote one of them or hire someone locally to be a manager to help run BLRSV.

Upon my return to NJ, I reported the dismal state of BLRSV to Brinkman and Netravali. Brinkman said that we should spend one year trying to turn around BLRSV. At that time, we would review the status of BLRSV and then decide how to proceed further.

Unfortunately, BLRSV was not going to get the time to be nursed back to health. Lucent had to tighten its financial belt in early 2001 and the Bell Labs budget was going to be affected adversely. Consequently, Bell Labs also needed to tighten its belt. BLRSV was going to cost Bell Labs several million dollars to run in 2001. Besides the cost, substantial effort was going to be needed to put BLRSV back on track. Moreover, since BLRSV had not yet integrated with Bell Labs and the rest of Lucent, the easy scenario was to close it and save money. BLRSV, which was opened in July 1, 1998, was quietly closed on February 6, 2001. This was the first ever closing of a Bell Labs Research facility.

ENDNOTES
ELEVATORS AND VIEWGRAPHS

As part of my one-on-one meetings with the BLRSV staff, I went to downtown Palo Alto with one of the researchers, whom I will refer to as Jake. As we were returning to BLRSV in the car, I decided to make use of the remaining fifteen minutes to learn about the Jake's research project:

> **Narain:** Jake, Please tell me about your research project. Describe the project and tell me what you are trying to accomplish.

> **Jake:** I can't tell you about my research project in fifteen minutes!

> **Narain:** Why is that?

> **Jake:** It will take much longer than that to explain the research that I have done.

> **Narain:** Why don't you give me an "elevator description" of your research project?
>
> **Jake:** What is an elevator description?
>
> **Narain:** An elevator description of a project is a brief summary of the project focusing on its salient characteristics.
>
> **Jake:** I am not prepared to give you, off the cuff, such a description of my research project.

I was taken aback. Why was Jake not telling me about his research? A warning flag had been raised.

The next day, continuing my one-hour meetings, I met with another researcher, whom I will call Sarah, who was working with Jake. After the usual pleasantries, it was time to move on to business. Here is how the conversation went:

> **Narain:** Please tell me about your research project.
>
> **Sarah:** There are only forty-five minutes remaining in our meeting. I can't tell you about my project in forty-five minutes. It will take many hours to do this.
>
> **Narain:** Why can't you tell me about the project in forty-five minutes?
>
> **Sarah:** You are new to my research's technical area.
>
> **Narain:** I may not be Nobel Prize material, but I think I am reasonably smart. Why don't you explain your research to me, the best you can in the remaining time?

Sarah declined to explain her research project saying that she needed more time to explain the project and wanted me to allocate one whole day. By now, I was convinced that I was not being given details about the project because there was not much going on.

On further probing around, I was soon to learn that the project existed only on viewgraphs. The researchers needed a large sum of money to proceed with their project, money that they had been hoping to get for some time now. I was going to work with the team to help them move forward one way or another, but BLRSV ran out of time.

8 Maps On Us

Driving (Not Research) Directions!

I AM EMBARRASSED!

INSTRUCTION	GO
Head NORTH-EAST on MOUNTAIN AV	1.4 miles
CONTINUE onto ASHLAND RD	1.1 miles
TURN LEFT onto LARNED RD	0.2 miles
TURN RIGHT onto PROSPECT ST	0.1 miles
TURN LEFT onto TULIP ST	0.2 miles
TURN RIGHT onto SPRINGFIELD AV	0.1 miles
...	...

I COULD NOT believe it! According to Columbus, our on-board navigation system, there was a shorter way for me to go home from work – 3.9 miles instead of the 4.1 miles it took me to go home the usual way.

I had been driving between my home in Summit and my office in Murray Hill for almost 16 years, 50 weeks a year, six round trips a week on the average (I would often go to work on weekends). I thought that I had figured out the shortest way. Had I known the shorter way, I would have saved about 1900 miles of driving over 16 years:

16 years × 50 × 6 trips to work/week × 2 × 0.2
= 1926 miles

Columbus found a shorter route than the one I used because, unlike me, it had all the road data. It had been programmed to find different types of routes such as the shortest and fastest routes. In my case, Columbus opted to make a left turn off Ashland Road. This turn was not an obvious one to make. The more natural turn was a few blocks down at a flashing yellow light, where I would slow down, and then turn left.

However, Columbus did not know one important fact – Larned Road is unpaved. Our road data did not give Columbus this information. Had Columbus known this, it would have told me to go a little further on Ashland, and make a left turn directly on to Tulip Street. This route would still be about 3.9 miles long, perhaps a few yards more than the previous one. As before, the left turn on to Tulip is not an obvious one to make.

But I am getting ahead of myself. How did we end up building an on-board navigation system? What did Columbus have to do with AT&T's telecommunication business?

FAT MINUTES

Early in 1994, AT&T was losing customers (both wireline and wireless) to other long-distance carriers who were aggressively courting and successfully luring them away with "sign up" checks and extremely competitive pricing. In addition, it was getting quite expensive for AT&T to lure customers away from other carriers.

Arun Netravali, then a senior manager, felt that if AT&T could use its network to offer new and unique value-added services, then customers would bond with AT&T and this would reduce customer "churn." If a customer left AT&T for another carrier, they would not be able to get AT&T's unique services. This would ensure that they were less likely to be tempted to leave AT&T simply because of the lower prices or the "sign up" checks offered by the competition. The new and unique services would convert commodity minutes on the telephone network to value-added or "fat" minutes that would not only help retain customers, but would also bring in higher revenues.

To buy McCaw Cellular, AT&T had paid a large amount of money, $11.5 billion dollars to be precise (based on the AT&T's stock price on September 19, 1994, the official merger date). Therefore, there was a strong desire in AT&T to increase wireless revenues substantially. Netravali wanted Bell Labs to help by inventing applications that made the wireless minutes "fat."

Netravali thought that an on-board navigation system for automobiles would be an interesting research project that would also help AT&T's wireless division offer a new service to its users. The on-board navigation system would offer maps, routes, and local (to the car location) business information such as hotel room availability and gas station distance and timings.

The map data was static and would be kept locally in the car. The business information was dynamic, i.e., constantly being changed, and would therefore be kept on a server in some data center. Getting business information would therefore require connecting to a server using a cellular (wireless) connection. An on-board navigation system, if sold by AT&T, would increase wireless traffic and reduce customer churn. And, of course, AT&T would generate additional revenues by selling the equipment for the on-board navigation system and the new service that would provide the latest business information.

RESEARCH (NOT DRIVING) DIRECTIONS!

Because Bell Labs culture allowed researchers to select their own research projects, any violation of this freedom was strongly resented. Indeed, I had been instrumental, at times with a colleague, for starting all the research projects that I had worked upon until now. In my sixteen years to that time, no manager had ever asked me to work on a specific project. They had, of course, their own opinions on the things I had worked on. And it was the same scenario with most of my colleagues.

Most research projects did not have immediate applications, if any, in AT&T businesses. Even the UNIX system and the programming languages C and C++ fell into this category. They did have a large impact on the academic and business worlds, but even they did not make AT&T any money. Two of my projects, the parallel programming language Concurrent C/C++ and the Ode object database also had an impact on the academic world – but not of the same order as UNIX, C, or C++. Concurrent C/C++ and Ode never became products.

After management began encouraging researchers to work on projects that they thought would address company needs, Netravali asked two researchers to start work on building an on-board navigation system. Although they did not say no to Netravali, they did not really want to build a system. They correctly surmised that building an on-board navigation system was going to be primarily development with little by way of research. They made some viewgraphs and started to look into the theoretical issues involved in building an on-board navigation system. Netravali saw that they were not enthusiastic about the project and that they were not approaching it the right way (focusing on system building).

Being keen on helping AT&T's wireless business unit, Netravali was eager to start the on-board navigation project. Eventually, he asked me if I would take on the challenge. I was agreeable to explore starting the project provided Netravali would commit to facilitating the business aspects of the project once we built it. Too many times, I had ended up with projects that a business unit was not interested in or for which AT&T did not have appropriate marketing and sales channels. Netravali assured me that he would help make this a business success. He advised me not to worry about the business issues, but instead to focus on building a real on-board navigation system.

I found the combination of mapping, driving directions, and cellular telephony to be very intriguing since I would be learning about several different technologies. I was not worried about the research issues at the start of a systems project. Time and again, having started to build something new and novel, we had come up with technical challenges that led to research publications and patents.

The on-board navigation project was my first case of "directed" research, one where my management told me what to do. I had no problems with this since I was eager to do something that could make money for AT&T. I was optimistic that the on-board navigation project had a good chance of becoming an AT&T product since Netravali was behind it. As it turns out, I was too optimistic.

COLUMBUS

Building the on-board navigation system would require resources. Fortunately, I had some resources at my disposal since I had been promoted to a manager a few months earlier. I was in charge of the newly created Database Systems Research Department. I had about 20 people in the department, mostly world-class researchers. My challenge was to convince a couple of them to work on the on-board navigation project. The operative word was "convince" since Bell Labs culture did not allow "ordering" researchers to work on a specific project.

For the on-board navigation project, I needed one or two system builders. Most of my researchers did not have experience building commercial quality software systems. Bill Roome was one of the exceptions. Having Bill Roome on the team would be a real asset because I wanted the on-board navigation system that we would build to be usable in the field. I was fortunate enough to be able to convince Bill Roome to join the project. That itself was a good indication that this project had technical challenges because otherwise Roome would have simply said no to me. Other members joined later.

We named our on-board navigation system Columbus in honor of the Italian explorer who discovered America. Columbus' goal was to reach India by going west from Europe, but he was not sure that this could be done. Similarly, our goal was to build a usable system that would tell people how to go from one place to another, but we were not sure if the quality of our road data would allow us to do this.

We rationalized our selection of the name Columbus by explaining to our management that it stood for

Communicating with and *Locating*, and *Updating Mo*B*ile* U*S*ers

Columbus would offer its users location related services such as maps and turn-by-turn driving instructions that could be printed or heard by the driver while driving. Users would also be able to ask queries such as

find gas stations within 5 miles of the current route

or

find sports stores in Morristown, NJ

and then get directions on how to get there, say from the user's current location, the shortest way.

The Columbus system would consist of several components. The user would have a device in the car for displaying maps, directions, and yellow page information. We called this device a personal travel assistant (PTA). The PTA would be the device that a driver would use to get maps and routes, and ask queries when on the road. Before traveling, a user could print maps and routes using his or her computer at home or in the office. The PTA would have a text-to-speech system that would read out directions as the driver progressed towards his or her destination. Each car would be fitted with a global positioning system (GPS) receiver to determine the location of the car.

We would keep some static data in the PTA, but data that was changing constantly or taking up too much space and used rarely would be kept on the Columbus applications server. The server would provide dynamic business information such as the current hotel availability, availability of movie tickets, and current weather information. The server would ensure that it had the latest business information by interacting with the servers of various businesses and as a result of the updates made by us.

The PTA could be used in both "at home or office" mode (high bandwidth, large display) and "on the road" mode (low bandwidth, small display). In the "at home or office" mode, the PTA would be connected directly to the office LAN and through it to the Columbus server or it would be connected via a relatively high bandwidth connection directly to the server. In the "on the road" mode, the PTA would be connected to the Columbus server by a low-bandwidth data connection setup over a wireless cellular connection.

The cellular aspect of Columbus is what would generate the "fat" minutes for AT&T. Of course, AT&T would also be able to generate revenues by selling the Columbus system itself. The potential market was huge, exactly what AT&T business folks liked. There were tens of millions of potential customers. Moreover, AT&T could make large deals with the automobile manufacturers.

The few on-board navigation systems available at the time typically stored all map and points of interest data locally. As a result, they could not offer services that depended on data that was changing all the time. Moreover, existing systems could not store the entire map and routing data for the USA in the navigation system because there was too much data. Because of our design, we could keep the rarely used map and routing data on the Columbus server.

To build Columbus, we needed several items that were beyond the scope of what we could do or had the technology to do. First, we needed map and route data. We could use the free map data, the Tiger data that was available from the US Census Bureau. Unfortunately, the data was not very accurate and it did not have the information that we needed to give accurate turn-by-turn directions. However, two commercial companies were offering much better data for a price. One of them, Etak, had taken the Tiger data and put a lot of effort into "cleaning" it and adding additional information, and it was continuing to do so on an ongoing basis. Etak also offered yellow page data (through a partner). We eventually decided to license the data from Etak, then a division of News Corp. that was eventually purchased by Sony. The fact that they had New Jersey data available immediately tilted the decision in their favor.

Next, we needed a GPS receiver, which was easily available in the market. A GPS receiver uses satellite signals to compute its location and velocity. It works with the GPS satellite system (funded by the US Department of Defense), which consists of 24 satellites, each of which orbits the earth in twelve hours. Between five and eight satellites are visible to a GPS receiver from any point on the earth. A GPS receiver requires signals from four different satellites to its compute location and velocity.

We needed to use the public cellular network for communication between the PTA and the Columbus server. At the time, public wireless data networks did not exist. A colleague and his team agreed to work with us to design and implement a new protocol that would allow us to transmit data over a cellular connection.

To store the yellow page data, we would use an object-oriented database, then a new type of database that stored data as objects as opposed to tables with rows and columns as done in the popular relational databases. Fortunately, we had just built an object database, the Ode object database, which was of commercial quality. The object database was not used in the initial version of Columbus, but it was used in a later derivative of Columbus called Maps On Us. (To be precise, we used its storage manager).

Finally, we needed a text-to-speech system that would speak out the text directions for the driver as he or she was driving. The directions would be given to the driver in steps based on the progress of the car. Fortunately, Bell Labs speech researchers had built such a system and we would be able to use it for Columbus.

We felt that the two risky issues were the quality of the road data and the accuracy of the GPS system (100 meters). We were not sure how these two factors would affect the quality of the road directions that would be produced by Columbus.

As the PTA, the device to be used in the car, we used a laptop computer as an interim measure. We did not envision its use in the commercial system that would be sold to consumers. With a laptop, the keyboard had to be used to enter information. Using the keyboard while driving was both difficult and dangerous. We envisioned the use of a speech recognition system for data entry when the car was in motion. Also, the laptop monitor was not the most convenient display to see when driving. We envisioned a display mounted on the dashboard.

But the laptop was a good starting point for the Columbus prototype. We expected the driver to plan a route using Columbus either before starting or by stopping somewhere or by asking the passenger (if any) to do so when the car was in motion.

The Columbus system was built with "2.5" persons working for about a year. To test Columbus, Roome would let Columbus take him wherever he had to go. I also went with him on several test runs. It was on one of these trips that Columbus showed me the shorter way home.

Once a route was planned, Columbus would "speak" out the directions, using the text-to-speech system, as the car moved towards the planned destination. A few hundred yards before a turn or before a road changed its name, Columbus would tell the driver what the next step was. (In case of a road name change, Columbus would say "Continue on" the road giving its new name.) If the driver went off the route planned by Columbus, say by making a wrong turn or taking the wrong branch of a fork, Columbus would soon realize the driver error and inform the driver. It would then ask the driver if it should re-plan the route.

Columbus could also be used to find services within a specified radius of the car's current location. For example, one could find the location of the nearest gas station and get directions to it.

While building Columbus we encountered interesting challenges from a research perspective that led to interesting database issues and a research paper. Suppose, for example, that a user makes a query to find the gas stations within a radius of 5 miles. The user then repeats the same query say after 1 mile. Even though the two queries are identical, the location of the car is different in the two cases. The gas stations returned by the second query would have some gas stations in common with the results of the first query because the circles defined by the query radii would overlap. To minimize the data returned by the second query, we came up with a scheme in which only the new gas stations were actually sent back to the laptop (remember we were doing it over the cellular network and at the time the effective bandwidth was low). The PTA would combine the new set of gas stations with the applicable gas stations from the first query to produce the result of the second query. Although we did not get around to implementing this scheme, a generalization of the above problem did lead to a research publication.

We gave demos of the system to many senior managers and they all pronounced that the system was impressive and did work as we had advertised. The question facing me was what next to do with Columbus. From a research perspective, we had accomplished the task we had set out to do. Now came the hard part – to move Columbus towards a business path.

COLUMBUS AS AN AT&T OFFERING

Although Columbus was started at the behest of Netravali, when it came time to connect Columbus with an AT&T business unit, we were essentially on our own. Netravali did help a bit, but I had to do most of the work. I had suspected all along that he would put the responsibility on me, but that was okay. Netravali had lots of things on his plate and some of them were probably more important and urgent than Columbus.

Unfortunately, there was no clear path towards making Columbus an AT&T business offering. In the early 1990s, researchers were being encouraged to collaborate with the business units and, at the same time, business units were being encouraged to work with Bell Labs Research with the goal of coming up with new products. This encouragement did increase interactions between Research and business units. We talked to many business colleagues and gave many presentations. However, it was not clear which one of these business interactions would eventually bear fruit with respect to Columbus. It was critical to find this out as soon as possible, since otherwise months could go by in a fruitless endeavor to collaborate with the business units.

To weed out business colleagues who were not genuinely interested in working with us, I would try to gauge their level of interest early on in our interactions. If the business units wanted to work with us, they had to commit some resources, for example, to do things such as market analysis, requirements specification, and system development. If a business unit was willing to do this, I was even willing to contribute more resources to a joint project because their willingness to commit of resources was a clear indication of their seriousness. The amount of the resources the business unit could or would contribute was not important. They could contribute a small amount as long as they demonstrated their commitment. Without such a commitment, there was not much point in working with the business units because they would have nothing at stake. With this "litmus test," it was relatively easy to prune out the non-serious business unit staffers who had an agenda different than a meaningful collaboration with Research.

As it turned out, none of the business colleagues we were having discussions with were seriously interested in collaborating with us to market the Columbus on-board navigation system. There were a couple of reasons for this. First, on-board navigation was a new product offering for which AT&T did not have appropriate marketing and sales channels vis-à-vis the automobile manufacturers or the car rental companies. Also, product management in the business units was typically not interested in new product or service offerings that did not have the potential of generating in the neighborhood of at least one hundred million dollars in revenues in the first year.

Without business unit interest, we were not in a position to move Columbus toward becoming an AT&T product. This was very frustrating, but at least we had built a very interesting system and learnt about several technologies that we would not have otherwise learned. However, I was not prepared to give up my effort to commercialize Columbus.

A NEW MEDIA IS EMERGING

By early 1995, I concluded that AT&T would not market Columbus as a product. It was time to salvage whatever we could from all the effort that we had put into building Columbus, understanding wireless and mapping. We had published one research paper, but, as far as I was concerned, this was not a very satisfying outcome of our efforts.

Around this time, the Web was starting to generate interest. Therefore, I suggested to Roome that it would be interesting to put Columbus on the Web. Roome's initial reaction was that Columbus would not work on the Web. Columbus was a highly interactive application that required it to be on the user's computer. Putting it on the Web would result in a slow connection to Columbus and a poor user interface. Columbus would be running on a remote server (who knows where). The user would have a low bandwidth connection to the server, and interface would be limited by the Web browser's capabilities. These were the days when the Internet connections were in the several Kilobit range (not megabits or even 56K) and there was no Java or JavaScript to facilitate the design of good and interactive user interfaces.

Despite Roome's initial negative reaction, I knew that he loved a challenge. In a few weeks, Roome came back to me and said that he had an initial Web version of Columbus working on the local LAN. I felt immensely satisfied since I had initiated work in a new and upcoming area.

By mid 1995, Roome had a working Web version of Columbus. We named it Magellan to distinguish it from the Columbus on-board navigation system.

ANOTHER BUSINESS OUTLET SPINS AWAY

We started using Magellan ourselves and found it quite useful. Of course, several naysayers kept telling us that no one would want to get turn-by-turn directions from the Web. But we had a gut feeling that they were wrong and we were right.

By early fall 1995, Magellan was working on the Bell Labs LAN with New Jersey and New York City metropolitan area road data. It getting about 1000 page views per day. Now Magellan was really becoming useful. One indication that we were on the right track was that our secretaries were using it and telling others about it. They found Magellan to be a valuable tool for giving directions to their bosses and to visitors coming to Murray Hill. Rather quickly, our user base started expanding by word of mouth.

Once again, as in the case of Columbus, the question was what to do with Magellan. How could Magellan make money for AT&T since AT&T was not in the Internet business? About this time, AT&T launched the AT&T WorldNet service, offering access to the Internet free for a few hours every month to its telephone customers and offering unlimited use for about $20 per month. I soon realized that AT&T WorldNet could potentially be interested in offering Magellan to their users as a service. AT&T WorldNet was a possible channel for Magellan and so it seemed for a while. I knew Jim Finucane who was responsible for the development of AT&T WorldNet service. He indicated an interest in Magellan.

However, fate seemed to be conspiring against us. On September 20, 1995, AT&T announced that it was splitting into three companies – AT&T, NCR, and a systems and equipment company, which would eventually be named Lucent. Bell Labs would go to Lucent, but about a quarter of the researchers would go to a new research lab, AT&T Labs, that would be created for AT&T. Within a few months after the announcement, I made up my mind to stay with Bell Labs, that is, with Lucent. AT&T would be keeping WorldNet, which was a service offering. Consequently, WorldNet and Magellan were going to end up in two separate companies. Since Magellan was a service, it was more suitable for AT&T, but it was going to stay with Bell Labs because of me.

There was still nearly a year to go before Lucent would spin off from AT&T. I again talked to Finucane about WorldNet's potential interest in Magellan, now that the company was splitting up and that I would be with Lucent. Finucane said that he was still interested in Magellan and that he could perhaps even give us development money, about $2 million or so, to make Magellan ready for commercial use. For this, Magellan would need to handle nationwide road data and yellow pages, be scaleable to handle millions of users, and be reliable enough to offer 24×7 service.

Unfortunately, over the next many months AT&T WorldNet's interest in Magellan never did move in the direction of becoming serious. By late spring 1996, a few months before Lucent was spun off, Finucane no longer seemed interested in Magellan and I was forced to conclude that AT&T WorldNet was not going to offer Magellan as a service to its users. AT&T managers had said that they wanted to offer value added applications on top of the AT&T WorldNet service and Magellan had seemed a nice fit. But AT&T had more important issues to address – the impact of the breaking up and dealing with its competitors.

It was time to go back to the drawing board once again.

FROM FRUSTRATION TO OPPORTUNITY

Not having a business outlet for our research was frustrating for all of us in Bell Labs who wanted to add value to the company. In the case of Magellan, the Magellan team and I were getting discouraged. No matter how hard we worked to build a system that people thought was innovative and represented leading edge technology, the chances of it going to the marketplace were almost non-existent. It was very unsatisfying to build yet another prototype that went no-where and that would eventually be put on the shelf. Our job seemed to keep repeating this process. I had done this several times by now and was quite frustrated.

Meanwhile, the Magellan user population had steadily been in-creasing by word of mouth. This growth was occurring even though only users inside the AT&T firewall could access Magellan and we did not have yellow page data. Moreover, this increase was occurring despite the fact that we had map and road data only for New Jersey and New York. If we got the data for all of USA and if we offered Magellan to the world, then based on extrapolating the Bell Labs population to the US Internet user population, we would see tremen-dous use of Magellan.

If no one in AT&T/Lucent wanted to take our research to the customer, we would do it ourselves. We would spin off a company from Lucent that would offer turn-by-turn directions on the Web. However, since at the time there was no formal process for starting a new business, I decided to approach Netravali. Over a lunch meeting with Netravali one afternoon, I told him about my frustration about the lack of business outlets for our work. I requested that I be allowed to take Magellan to the marketplace, that is, run it as a commercial website. I would need $600K for licensing the road data for commercial use, buying servers to run Magellan, leasing T1 lines to connect to the Internet, and hiring an additional programmer or two.

Besides enabling us to launch the website, getting the money would also signal management approval of the path I wanted to take. There were other important issues to be addressed such as making the business plan and hiring the staff for business development, marketing, and sales. However, I would address these topics later.

I had expected Netravali to say no. Much to my surprise, he said yes, I could have the money, subject to two conditions. First, I was to meet with Mel Cohen, research effectiveness vice president, and convince him about the viability of making Magellan a business. I also needed to make a business plan using Cohen's template. Second, I was to convince Barry Karafin about the business proposition for Magellan. Karafin was a consultant to Netravali. He had been the vice president of strategic planning and marketing in the part of Lucent that was eventually spun off as Avaya.

Cohen's focus was on hardware boxes that we could sell, not on Web services. He wanted to know what boxes we would sell or what software system we would license to users. Although very helpful in telling us about the process used in Bell Labs to create a business plan, Cohen was not sure about the proposition of launching Magellan as a business. He correctly viewed Magellan as purely a software service play in a new and unproven medium, one he did not know much about from a business perspective. He also did not believe that people would be interested in getting driving directions from the Web.[128] Nevertheless, I made a business plan starting with Cohen's template.

The next step for me was to try to convince Karafin about the viability of Magellan as a business. Meeting Karafin was an interesting experience for me. At the first meeting, he did not want to see a demo of Columbus. As a technical person, I wanted to show him what we had built. Karafin said that he was not interested in looking at the technology at this stage.

Karafin, as a first step, simply wanted to get a read on my commitment and passion. He did not tell me this, but I heard it from Netravali. Karafin wanted to understand the Magellan business model, that is, how Magellan would make money. I said that our model was to offer the service free to end-users and generate revenues in two ways. First, we would sell banner ads. The conventional wisdom in 1996 was that a website could be a profitable business with just banner ad revenues. The only issue was to ensure that enough users came to the website and generated the requisite number of page views. Second, we would charge companies for providing them maps and routes for their websites. Potential businesses that could be interested in maps and routes would be those in the travel and tourism industry, health care, etc.

This was the first time I had to justify what I was doing from a business perspective. I was nervous. Fortunately, Karafin found me to be very committed to making Magellan into a business and was reasonably satisfied with the business model. I had the green light from Karafin.

Karafin's support was strong enough for Netravali to give me the go ahead. Netravali also said that I should work with Lucent's New Ventures Group to ensure that I got proper business support.

GETTING READY FOR BUSINESS

The Magellan team members were very excited. We could potentially be spinning off as a separate business, a startup. By now, the team had grown to five persons. Besides me, there was Roome and three other staff members (all technical). This was an opportunity for us to convince skeptics that Magellan was a real system, which people would use, and which would generate revenue.

We set ourselves the goal of launching Magellan commercially by the end of the year. Roome felt this was doable. We had to pick a new name for Magellan because another company had trademarked the name Magellan. Available good domain names had already started getting hard to get. Therefore, we had to be creative in coming up with a good name for which a domain name was still available. All of us liked the name Maps R Us that we came up with, but we felt that Toys R Us would give us a tough time because of the similarity in the names. Roome suggested several other names including MapBook, MapBiz, Maps On Us, and Way2Go. We eventually picked Maps On Us, a variation of Maps R Us that we all liked (or did not oppose). We eventually came up with the following logo:

While we were heading towards trying to spin off Maps On Us as a separate company, preparations were taking place for spinning off Lucent out of AT&T. In addition, Netravali had become the vice president of research.

We started to work furiously to get Maps On Us ready by the end of the year. But there were several impediments that needed to be overcome. Creating a business was uncharted territory for Research. Although I had Netravali's support, I did not have the full support of my more immediate management, who were bound to the traditional Bell Labs culture. For example, one manager in my line of management, Ravi Sethi, who was at the time the research senior vice president, felt that it was time to move past Maps On Us since it was not central to Lucent's core business.

I explained to Sethi the origins of Maps On Us, how it had evolved from the Columbus on-board navigation system that we had built for helping the AT&T business. Our goal had been to build an innovative service to help AT&T Wireless offer "fat" minutes to reduce customer churn. When this avenue did not work out, we evolved Columbus to operate on the Web, to provide AT&T WorldNet with a new service offering. WorldNet had been interested, but this avenue for our work vanished with the AT&T trivestiture. I was convinced that we had built a very useful system and our goal was to monetize it for Lucent. Lacking appropriate business channels in Lucent, launching Maps On Us as a separate company was one way of getting a return on investment for Lucent. Otherwise, Maps On Us, like many other research projects, would be put on the shelf. Another reason I wanted to move forward was researcher frustration. Roome and I wanted to make a difference with what we had built. Researchers who could build innovative systems were in great demand outside of Bell Labs and, if the systems they built went nowhere, they would leave Bell Labs. Therefore, it was important for us in management to explore all possible avenues that could lead to success.

Sethi appreciated my thinking and agreed that we should continue with Maps On Us. I was determined to try the new channel (for Research) of launching a company to offer Maps On Us as a service.[129]

LUCENT NEW VENTURES GROUP

Around the time we were thinking of launching Maps On Us as a separate company, Lucent was in the process of creating an internal venture organization. Lucent New Ventures Group (NVG) was formed circa 1996, soon after Lucent was spun off from AT&T. The goal was to create a world-class internal venture organization that would help unleash the powerful technologies created by the Bell

Labs researchers into the commercial arena. The bottom line of NVG was to get a return on investment for Lucent from the technologies developed by Bell Labs that were not central to the Lucent business and which were not of interest to the business units. NVG would help set up groups to commercialize these technologies and spin them off as separate companies.

Although Bell Labs Research was shifting to help the business units, many research projects were still not central to the core business of Lucent. Moreover, as the parent company kept getting smaller and smaller because of events such as the AT&T divestiture and trivestiture, the core business kept shrinking and along with it the chances of a research project being relevant to the company.

Eventually, research projects developing technologies that are of no use to the parent company are stopped and the researchers move on to another project, and the technologies and systems developed forgotten. NVG was to take such technologies and systems and help create new ventures around them. One or more researchers were expected to transfer to the new ventures. NVG would provide business support, help form a management team, invest in the venture, and help raise money for the venture. For its efforts, NVG would get an ownership in the venture.

Researchers welcomed the formation of NVG, which they hoped would bring with it business expertise, connections to venture capitalists, and experience in raising money. More importantly, NVG would give the researchers another channel that would allow them to take their work to the marketplace. If the Lucent business units were not interested in a particular technology, then the researchers, the ones who created the technology, could work with NVG to explore the creation of a new business.

At Netravali's behest, I enlisted NVG to help Maps On Us. Our team lacked business expertise and any business help provided by NVG would be welcome.

PUTTING THE TEAM TOGETHER

The core Maps On Us technical team was in place by early 1996. Organizationally, the Maps On Us team was in Research, just because each of us was in Research. Missing was the marketing and sales staff. I could not hire such staff in Research simply because Research did not hire staff with such a skill set. This is where NVG should have stepped in to hire marketing and sales support, but they did not. Unfortunately, they were also in the early stages of setting up business.

Fortunately, Maurice Kuritz, a business manager in the internal Lucent venture named Inferno, had been asked by Netravali, through Sethi, to help develop the Maps On Us business plan. It was clear to Kuritz that Maps On Us needed marketing and sales staff or it would go nowhere. At his own initiative, Kuritz took on the responsibility of helping Maps On Us hire two marketing/sales persons and placed them in the Inferno organization. Kuritz was able to use some funds that had been allocated to help potential startups. Without Kuritz's help, Maps On Us would never have operated in the commercial world.

Maps On Us now had a team of seven in place. Beside myself, there were four technical staff and two marketing/sales staff. Maps On Us was ready to go.

LAUNCH

With the basic team assembled in the summer of 1996, we started counting down to the launch date, which we had set for the end of 1996. There was a lot of work for our small team to do in six months.

Besides getting the Maps On Us software ready, there were many other things to do. We had to order the servers, the racks, and the disks to store the large amount of map and road data and yellow page data. We had to connect these machines and place them "logically" outside the Bell Labs firewall, which protects the company LAN and resources from unwanted intruders. Amongst other things, the Bell Labs firewall does not allow users from the outside to access Bell

Labs machines inside the firewall. However, we did need to allow the "external" machines to communicate with the machines inside the firewall because we would be periodically updating the Maps On Us software and data. In addition, we had to bring the Web logs on the external servers inside the firewall. An interesting point to note is that our computers, both inside and outside the firewall, were sitting next to each other in the computer room. In any case, we had to devise a safe strategy that would allow the internal machines to "talk" to the external machines. Much of this is routine now, but in 1996, we were on the frontiers of Web technology.

Since Maps On Us would have many servers to handle the large number of users we were expecting, we would have to design and implement a load distribution strategy. Next, since we were a very small team, we could not monitor the systems 24 × 7 by being physically present. Consequently, we had to design and implement a strategy to handle machine failures automatically without any interruption of service. In addition, even though failure handling was going to be automated, we needed to design and implement a pager plus email notification system to alert the key people whenever there was a problem with the system. Although Maps On Us might still be running, we needed to understand what was happening. For example, if one of the servers crashed, our automated system would ensure that the other servers would take over its load. However, we wanted to find out the cause of the problem and see if the situation required immediate attention or if it could wait until regular work hours. Software to do much of this is now available in the market, but in 1996 we had to design and implement it ourselves.

Although it would take time and effort, we knew how to do the technical stuff. We also had to do marketing and sales, something that was very new for us. Marketing and sales material had to be prepared, banner ads had to be designed, an ad rep firm (to sell banner ads) had to be selected, and so forth.

I worked with the Maps On Us team in every aspect that I could. I even spent time in the computer room helping to get the equipment ready. Many people were surprised that a senior manager like me was working with the Maps On Us staff in the computer room helping to set up equipment. To me, this did not seem unusual. Anyone running a business, particularly a small business, should do whatever it takes to make the business successful.

We were (or at least I was) very confident that Maps On Us would get many users. We hoped that this would lead to ad revenues (the first part of our revenue model). However, I was not confident about personally being able to sell our services to enterprises (the second part of our revenue model). In my nearly two decades at Bell Labs, I never had the opportunity to meet a real AT&T or Lucent customer. And, of course, I had never sold a product!

Most of all, I was worried about sullying the Lucent brand with a Maps On Us that did not stand up to commercial use and quality expectations. We had never built and operated a service that ran 24 × 7. A whole host of things could go wrong. The servers could crash, the software could have bugs, our driving directions could be absurd, etc. Lucent had spent a lot of money establishing the Lucent brand. Any negative publicity arising from a shoddy Maps On Us service would certainly, at the very least, lead Lucent to close Maps On Us.

Maps On Us was launched without much fanfare on January 13, 1997:

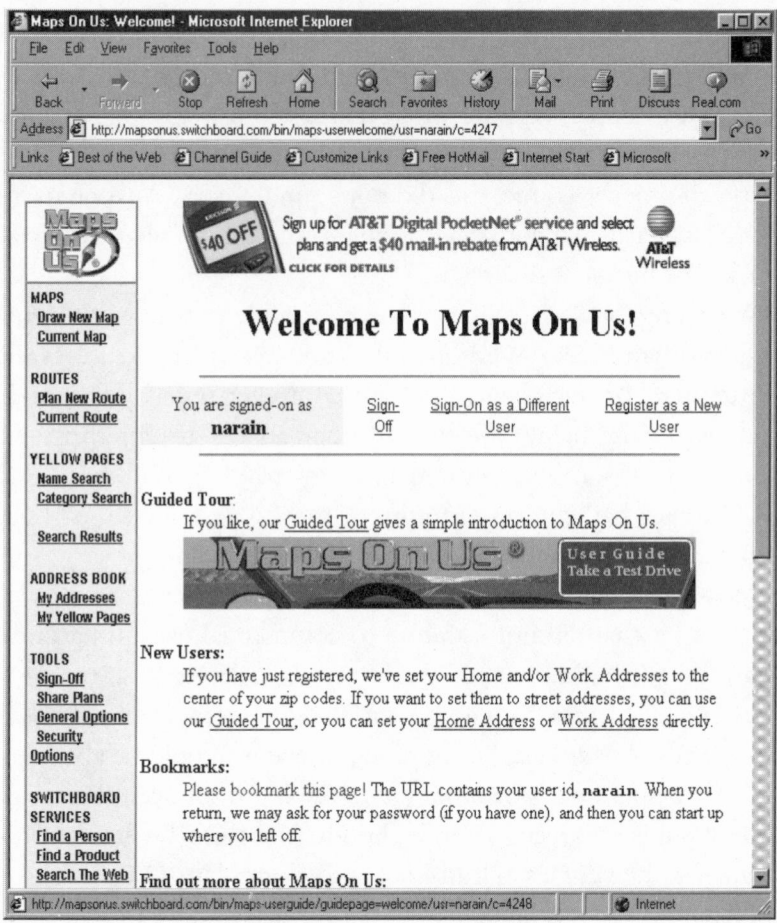

Maps On Us was supposed to be launched as a new venture. We had hoped that we would shift out of the Research location so that the team could focus on the business. However, it just seemed easier to continue to operate out of Bell Labs Research in Murray Hill, which is what we did.

In retrospect, it would have been better for us to move out of Research, both from a team and a customer perspective. This was partly my fault. I did not push the issue strongly because there were other things that I was more worried about. Our offices were the same as before and they were intermingled with the offices of the other researchers who were working normal working hours solving research problems. We, on the other hand, were committed to providing an innovative 24 × 7 service and were working long hours including weekends to make this happen. Some team members complained that, unlike the other researchers, they were putting in a huge number of extra hours, but they were getting nothing extra in return. Specifically, none us had been given equity in Maps On Us as we had expected. We were expecting to get phantom stock in Maps On Us. The phantom stock would become real equity when Maps On Us was launched as a separate company. I hoped that this issue would be resolved soon. As it turns out, we ended up not getting any equity in Maps On Us.

From a customer perspective, it would have been nice for Maps On Us to have its own offices because when customers came to visit us, the huge Bell Labs complex would overwhelm them. They were not sure if we were a bunch of researchers or if we were a team committed to providing them with 24 × 7 service. We looked more like a Bell Labs project than a company.

USING MAPS ON US

Maps On Us provides users with three main capabilities – maps, turn-by-turn driving directions along with a map of the route, and yellow page searches, the results of which can be used to draw maps or get driving directions. Each of these capabilities has many features. For example, when requesting driving directions, a user can specify the use or non-use of highways or ask for the shortest route or the fastest route.

Maps On Us produces clear and elegant maps of any user specified location. The location can be a standard postal street address, an intersection of two streets in a city, just the city and state, just state, or just the zip code. In the last three cases, the map is centered within the area specified. For example, the following is map of the area around the Bell Labs facility at Murray Hill:

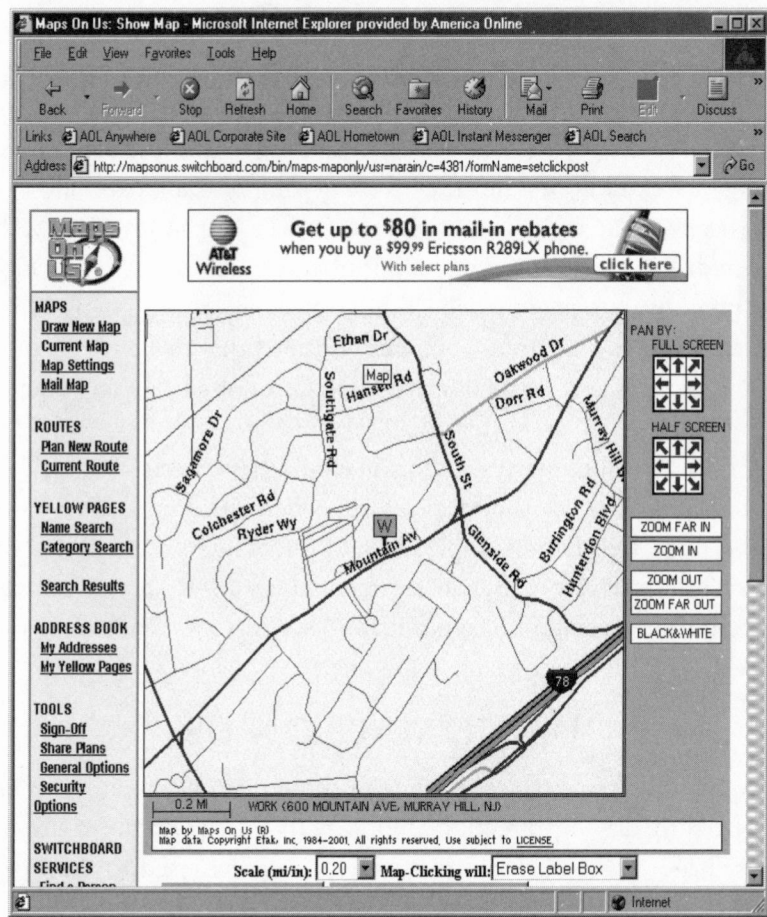

The marker with the label "W" specifies my (former) work address, the address of the Bell Labs complex at Murray Hill. I preset this when registering to use Maps On Us.

A user can do many things with the map produced by Maps On Us. For example, the user can zoom in or out, label map features, and see latitude and longitude.

To get turn-by-turn driving directions, a user specifies the start and end addresses of the route and selects the option specifying the preferred type of driving directions (fastest, shortest, preferring highways, avoiding highways). After this, all the user needs to do is click to plan the route. Maps On Us returns a wealth of information for the traveler, for example, the distance to be traveled, the expected travel time, a map with the route plotted on it, turn-by-turn directions for the route, and optionally, detailed maps for each one of the turns.

The following images illustrate the map of a route and the accompanying text directions for going from Bell Labs at Murray Hill, NJ to Newark Airport. First the map of the route:

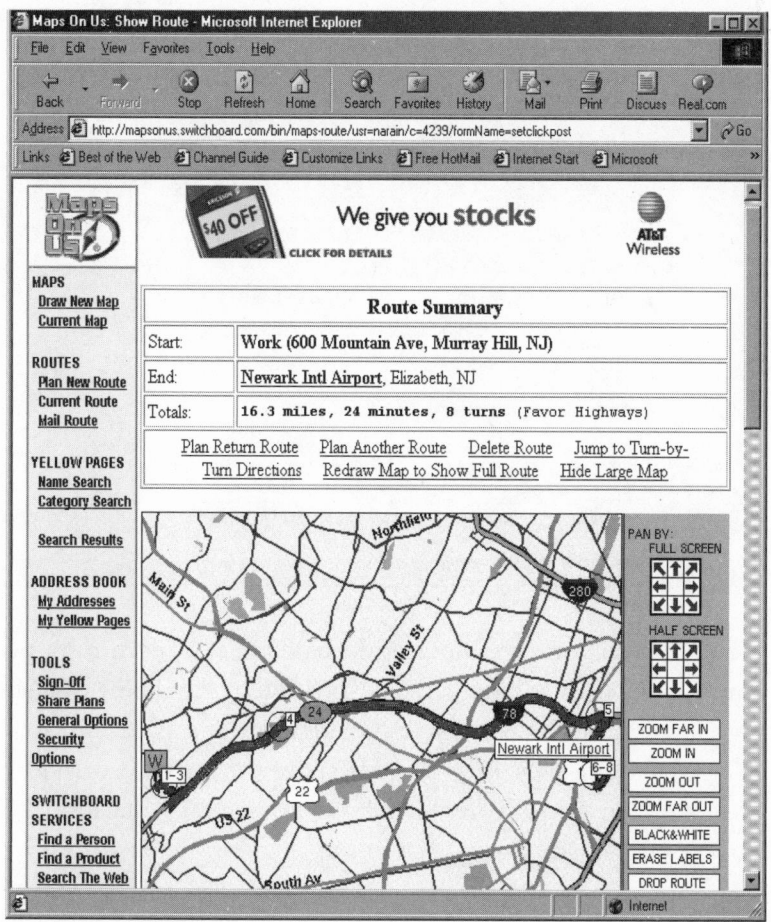

Now the text directions:

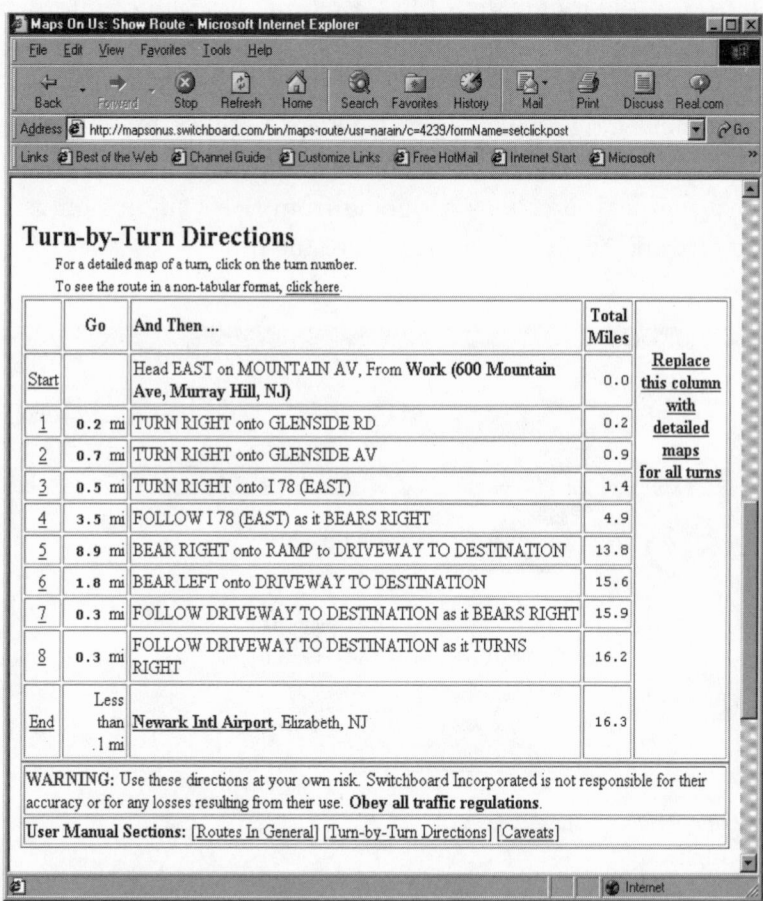

Maps On Us can give the user detailed maps of each turn, plan a return route, specify intermediate destinations, plan a best (the shortest) route when more than two destinations are specified, and so on.

The Maps On Us yellow page database can be used to find contact information about commercial, governmental, educational, and other types of organizations. For example, it can be used to find the address and telephone number of Newark Airport or Chen's Chinese Restaurant in New Providence, NJ.

The following Web page shows the start of a search for Newark Airport:

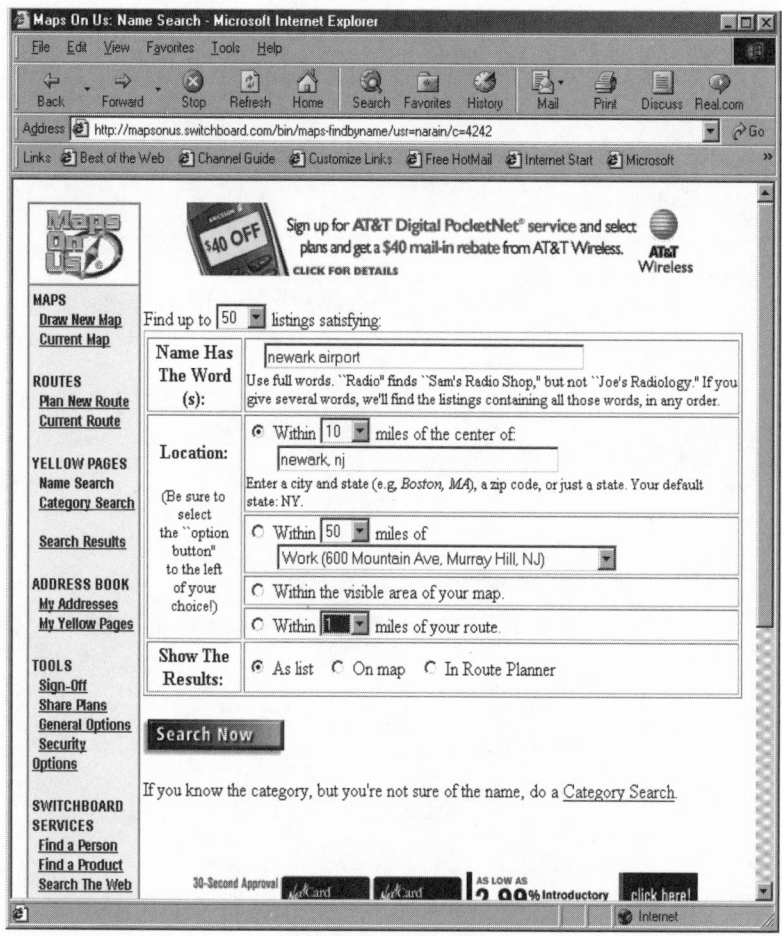

Yellow page searches can be done by specifying words in the name of a business or an organization (as shown above) or in a business category. A user can specify the radius of the area within which Maps On Us should do its search.

Unlike other yellow page databases, the Maps On Us database is integrated with the maps and route planning facilities. So having found the business or organization of interest, say Chen's restaurant, Maps On Us can be used to get a map of New Providence or to get directions on how to go to Chen's restaurant.

The search shown above produces the following results that can be used to plan a route from Bell Labs, Murray Hill, NJ to Newark Airport:

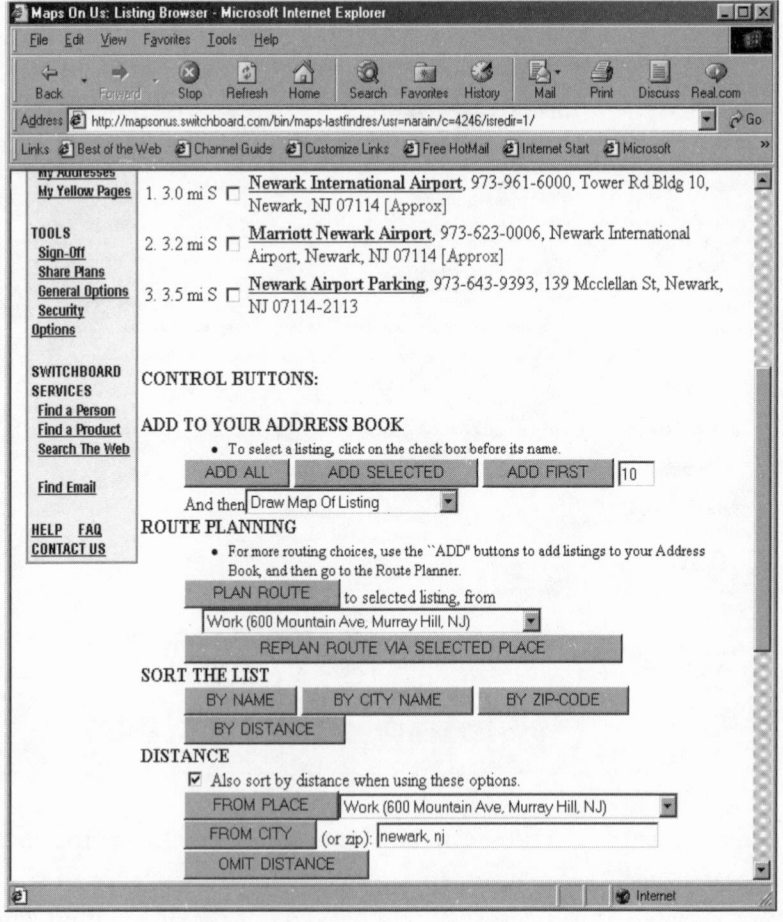

Selecting Newark Airport and then using the PLAN ROUTE option along with the previously stored address of Bell Labs generates a map of the route and turn-by-turn directions to Newark Airport.

USERS, APPLAUSE, AND AWARDS

Maps On Us got 50K page views on the first day of its operation as a commercial service. The number of Maps On Us users grew very rapidly. We started by getting our first non-Bell Labs users as a result of the Maps On Us launch announcement on the newswires. We also advertised a small amount by buying banner ads on other websites. However, most of our rapid increase in the number of users and in page views could be attributed to our winning awards or getting good reviews in the press. Each such event resulted in a quantum increase in the number of users and page views.

As a result, to keep up with the use of Maps On Us, we had to increase the number of servers and the bandwidth of our connection to the Internet.

Maps On Us received a tremendous amount of good press. The satisfaction of having our work appreciated in the national press was something none of us in the Maps On Us team had ever experienced before. In contrast, even though some of us had developed innovative systems and technologies before Maps On Us, the best appreciation we had ever received was a pat on the back from our management. In the academic research world, awards and honors generally have limited visibility. However, with Maps On Us, we could get nationwide, and even worldwide visibility. Our nationwide visibility, as a result of the acknowledgements in the press that we had produced an excellent system, was extremely gratifying.

Here are some extracts from the positive press received by Maps On Us:

> *But even after your plane lands, there's still the rental car and navigating your way through a strange town. That's where Maps On*

Us is a blessing from the gods. OK, it's really from Lucent Technologies, but no less a miracle. ... Best of all the maps are uncannily accurate. This is the only program to pass my personal test – airport to my house – and even knows my secret shortcut.

— Ron White, "Map Mania."
PC Computing Internet, October 1997.

There are dozens, perhaps scores, of Web sites that provide maps for business and pleasure travelers, and Maps On Us is the smoothest and perhaps most versatile that I have sampled.

— L. R. Shannon, "Planning with Online Maps."
The New York Times CyberTimes, September 21, 1997.

A street-savvy navigation tool that does everything but slice and dice.

— Top 10 Educational Reference Sites.
Interactive Week, August 11, 1997.

My favorite of all the programs I tested was Maps On Us (www.mapsonus.com) developed by Lucent Technologies.

— Paul Grimes, "Mapping it Out."
Chicago Tribune, June 1, 1997.

In addition to the positive press, Maps On Us also kept on getting awards. We were tremendously proud of what we achieved with a small team. Here is a partial list of the awards Maps On Us received (before the sale of Maps On Us to Banyan Systems in May 1998):

Award	Giver	Date
Editor's Choice	GOTO.com	January 9 1998
Times Pick	LA Times	October 21, 1997
Top 100 Website	PC Magazine	May 1997
Editor's Choice	LookSmart	April 1997
Site of the Week	Information Week	February 3, 1997
Site of the Day	PC Magazine	January 28, 1997
Hot Site	USA Today	January 16, 1997

NOT WITHOUT BLEMISHES

Maps On Us was an extremely well built system, but it was not perfect in giving road directions. This had nothing to do with our software development, but instead, the rare bad directions it gave had everything to do with the less than perfect road data that we had to use. Our maps and directions relied on the accuracy of the underlying road data, which was very good but not perfect. The problem was not specific to us but also affected our competitors.

Etak, the company from which we had licensed the road data, was continuously improving the accuracy of its data. Collecting and updating road data for the entire USA is an enormous job. The data is not static since all the time new roads have to be added and the status of some existing roads has to be updated. Consequently, some inaccuracy in the road data was to be expected.

The road data did not include many private roads such as those in developments. The road data also had data entry errors such as missing road segments (a road in the Etak data was represented as a series of connected segments), incorrect road names, and roads classified incorrectly allowing right turns on to one-way roads going left. The data was always going to be behind the current state of the roads in the USA since it did not reflect road changes that had taken place since the last data update. Etak supplied data updates quarterly.

However, road data for any particular region could be out of date by much more than three months since not every region was updated quarterly. The out-of-date data did not reflect new roads that had been opened, existing roads that had been closed, detours that had put in place, and two-way roads that had been made one way, and so on.

Inaccuracies in the Etak data would sometimes make Maps On Us produce interesting, actually absurd, directions. One particularly absurd set of directions was caused by a missing road segment near the entrance to the Holland Tunnel in Manhattan (New York City). The Holland Tunnel connects lower (downtown) Manhattan with New Jersey. Directions produced by Maps On Us for going uptown from some place in lower Manhattan, instead of having the form

Go North

followed the following pattern:

Go North to the Holland Tunnel
Take the Holland Tunnel to New Jersey
Go North in New Jersey
Take the Lincoln Tunnel back to Manhattan
Go North

We spotted this error since we were familiar with the Holland Tunnel and Manhattan. We fixed the problem by manually inserting the missing road segment into the data. In all likelihood, there were probably other road segments missing in the data that we never found.

Roome fixed many errors in the Etak data with a data improvement program that he wrote. This program was a rule-based system. For example, if two interstate highways appeared to be intersecting when one was simply going over another, the data improvement system would modify the data to not allow turns at the apparent intersection.

Early on, Maps On Us itself also needed some improvements. For example,

> *Maps On Us is distinguished from the rest by seven features. ... The advantages of Maps On Us extend beyond those standard features. ... For all its advantages, though, Maps On Us also has its flaws. In its current version, it has trouble handling complex hyphenated addresses like*

$$82\text{-}31\ 213^{th}\ Street$$

or towns with multiple names.

— Robert E. Calem,
"Navigating the Real World on the Web."
The New York Times CyberTimes, February 21, 1997.

We fixed the above problems quickly and we were always ready to improve our user interface.

REVENUES

Maps On Us had a two-pronged revenue model with revenues from banner ads and custom map services that we would provide to a few large customers.

Our revenues were based on our assumptions about the growth of the Web. We had considered three monthly growth scenarios: conservative, moderate, and aggressive:

	1997	1998	1999
Conservative	20%	6%	4%
Moderate	20%	8%	6%
Aggressive	20%	10%	8%

By the end of 1997, we were getting about 500K page views per day, a phenomenal growth rate since we started from less than 1K page views per day just about a year ago (just prior to its launch). We had assumed we would be starting with 25K page views a day and would end up with about 200K page views per day by the end of the year.

We assumed that we would get about $36 CPM (cost per thousand) for the banner ads that users would place on our website. We also assumed that we would be selling our services to a few large customers and expected about $250K in revenues in the first year.

We estimated that by the end of 1999, three years after starting, we would have generated revenues of $14 million in the conservative growth case, $18.5 million in the moderate growth case, and $25.3 million in the aggressive growth case. The corresponding profits would be $5.4 million, $8.7 million, and $13.5 million, respectively.

FIRST YEAR REALITY

Maps On Us started to generate revenues soon, but they were coming in more slowly than we had projected. We were not going to meet our first year revenue projections. We had been very optimistic in our projections. There was not much data available for Internet businesses, which we could have used to base our projections.

Several things went against us, preventing us from meeting our revenue projections. Both parts of our revenue model were affected. The conventional wisdom, prevalent at the time we made the business plan in 1996, was that if a website had page views, it would get ad revenue. Soon after we started, the buzz was that the top few sites, like Yahoo, were getting 80% of the ad revenues and the remaining huge number of sites were scrambling for the remaining 20% of ad revenues. Although within a few months we started getting significant page views, on the order of 200K per day, we were not one of the top few websites.

We had assumed that, since Maps On Us was owned by Lucent, we would be able to hire an ad rep firm to sell banner ads almost from day one. Unfortunately, it took us about two months to sign on an ad rep company – they wanted Maps On Us to have a couple of million page views a month before they would consider taking on the ad rep job.

The ad rep company we hired, Katz Millennium Media, did not deliver any revenues. We disengaged from them after a few months and then went to Petry Interactive (which later became 24/7 Media). Petry started delivering revenues very quickly, but we had lost several valuable months in our dealings with Katz.

Finally, the ad rates that we were getting were significantly lower than the $36 CPM we had anticipated, and we were not selling our full inventory. The average CPM across the whole site for us was in the single digits.

The second part of our business model called for selling mapping services (as a map application service provider) to a handful of large customers. In the first year, we expected to get about $250K from these customers. Our pricing was a fraction of what it would cost the customers to develop mapping services on their own. However, our business model for selling to large customers did not hold because customers did not believe the prices we wanted to charge them were worth the potential benefit. More importantly, competition was supplying mapping services at rates significantly lower than what we had planned (we had two primary competitors, Vicinity and MapQuest). After all, these were the days when a lot of material (both tangible and intangible) was being given away free or below cost. So we had to go back to the drawing board.

The right business model turned out to be many small customers paying around $15K per year, with some customers paying much more. Since we were getting many customers, we were expected to make more than the $250K we had expected to make by selling services in the first year.

We were having good success selling mapping services but not with ad revenues. As a result, we shifted our focus from ad revenues to revenues from selling maps and routes to customers who would offer it to their users. Our service revenue was an annuity. Assuming we did a good job, our customers were likely to stay with us, year after year. The cost of switching from Maps On Us to a competitor was significant enough to prevent a customer from leaving us unless we did a bad job. The same was true for the customers of our competitors. It was going to be hard to woo them away from the competition unless they also did a bad job.

We would have had more service revenue, but for the fact that one of our two marketing/sales persons was not good at sales. As the person who hired him, I have to take the blame for hiring a person with the wrong skill set. Although I can quickly figure out whether or not a technical person is right for the job, I did not have the experience to do the same for a marketing or sales person. To make a long story short, I was forced to replace him. Unfortunately, the replacement process took about two months, a long time in the short life of Maps On Us, which had been in operation for only about seven or eight months.

We had conservatively projected about $1.2 million as first year revenues (the number had been revised downward from about $1.6 million because of the scenario with ad revenues). Our projected expenses for the first year were about $1.8 million.

The revenue picture looked as follows (approximately):

o Q1: basically $0.
o Q2: about $50K.
o Q3: about $250K in signed deals.
o Q4: about $500K expected.

We were not going to meet our first year's revised revenue projections. Nevertheless, despite the fact that we had been slow in generating revenues, we had solved the problem with the ad rep firm, revenues were picking up, I had hired a new sales person, and we were optimistic that we were on track.

EUPHORIA TO GLOOM

The Maps On Us team was ecstatic with the growing number of users and the extremely positive press. We had lagged behind on generating revenues, but I was confident that we would break even in the coming year. Our starting problems were behind us and we were catching up on revenues. Moreover, we had many other pluses. Our expenses were low, about $1.8 million dollars per year, and not increasing much, our user base was increasing rapidly, and we were receiving a lot of recognition, which would make it easier for us to get new customers and more ad revenues.

However, Karafin, who had been continuing with Maps On Us as a consultant, was not happy with the state of Maps On Us. For example, Karafin felt that having many small customers instead of a few large customers would require much more resources. I explained to Karafin that we could handle the increased customer base without requiring additional resources since we had automated the delivery of map services and because we would not be offering custom services to the small customers.

Our revenue model was a service revenue model. Although we would get money up front for the Maps On Us service, we would not be able to book it upon receipt. We could book it as revenue as we delivered the service. The revenue generated by map services was booked, say monthly, as we delivered maps and routes. I felt that having service revenue was great since it was like an annuity. Karafin would have preferred revenue that could be booked immediately, as in the case of products. I was confident that we would renew most of the customer contracts (most of our customers, especially the large ones, were not dot-coms). However, Karafin said that we had no data about retaining customers. Unfortunately, despite my confidence, I had no data to show Karafin.

Karafin was justifiably concerned that we were behind in our revenues with respect to our business plan. We had some initial glitches, but we had solved them and were on our way. We were getting great press, our user base was increasing, our ad revenues were increasing, and we had started to get traction with customers.

The fact that we were behind in our revenues was the primary factor in our losing Netravali's backing of Maps On Us. To a lesser degree, Karafin's concerns about Maps On Us were also a factor. Unfortunately, I was not able to convince Netravali that we were back on track after some teething problems.

At a Maps On Us "board" meeting in the fall of 1997, attended by NVG, including its head Tom Uhlman, Netravali brought up the issue of service revenue. Despite signed contracts, the point of contention was that this revenue was not booked. The fact that it was an annuity seemingly did not matter. After the presentation, I was asked by Uhlman to leave the room so that the board could have a private discussion without me.

After the meeting, Steve Socolof and John Braskamp of NVG stopped by my office to tell me that NVG was going to install a CEO to head Maps On Us, essentially taking control away from me. They would offer me two CEO choices and I could select one. If I did not agree to this arrangement, Maps On Us would be closed.

I did not want someone else to run Maps On Us. After all, I had conceived of the idea, built a team, and created the business. We had been in business for about nine months, revenues had been delayed, but after some startup problems, we were now on track. Karafin had agreed with me that I had done essentially everything right, although some things had gone against me. Despite the revenue shortfall, I had done a good job. Our competitors were also suffering and we had done no worse. Since I was reluctant to let someone else run Maps On Us, Socolof wrote an email to Netravali and Uhlman recommending that Maps On Us be closed.

After some soul searching, in the interest of the Maps On Us team, I agreed to let NVG appoint a CEO to whom I would report as the president of Maps On Us. However, I agreed to NVG bringing in a CEO provided the new CEO was experienced in running a startup and had Internet business experience. Such experience would be key to making Maps On Us a successful business.

As it turned out, NVG appointed a CEO of their choice to run Maps On Us. The new CEO did not have the Internet business experience that we had agreed upon. However, I hoped that coming from a business background, the new CEO would be able to put our revenue growth on a fast track.

THE END APPROACHES!

Appointment of the CEO by NVG was the beginning of the end of Maps On Us (in Lucent). Starting from the very first meeting with the Maps On Us team, the CEO, being unfamiliar with the Research culture, was unhappy at the way researchers openly asked some difficult and probing questions. I also had disagreements with him and so did some of the Maps On Us team members on the operation of the team itself. Many of the disagreements were due to the cultural differences between our research background and the business unit background of the CEO. More importantly, on the business (revenue) front, things got worse for Maps On Us. As the situation on these two key fronts did not seem as if it was going to improve, and the fact that we had not been given any equity by NVG, the Maps On Us team was about to fall apart. Consequently, I recommended to NVG and to Netravali that we sell Maps On Us. The CEO was unhappy with my recommendation, but the research team was ready to quit.

NVG accepted my recommendation to sell Maps On Us. A few months later, we sold Maps On Us to SwitchBoard, which was then part of Banyan Systems.

Lucent Sells Mapping Unit
to
Banyan Systems

Lucent Technologies Inc., a maker of telephone equipment sold its Maps On Us Internet map unit to Banyan Systems, Inc.

— *The New York Times*, Page D1, May 19, 1998.

We did make some money with Maps On Us!

RETROSPECTIVE

Our attempt to spin off Maps On Us out of Lucent was an effort to address the channel issue by taking Maps On Us directly to the customers. Maps On Us was too small for any Lucent business unit to be interested, but more importantly, it was not in the core business of Lucent. Although, we did not succeed in spinning off Maps On Us as a separate business, we learned a little about creating and running a business, dealing with customers, and several new technologies. With Columbus, we learned about GPS, text-to-speech, and mapping. With Maps On Us, we learned about the Web early in its development and were able to apply our expertise to other research projects, for example, Web-based customer care.

Providing a service to tens of thousands of users who used Maps On Us daily was an exhilarating experience. Working long hours seven days a week was fun because Maps On Us was making a difference. The enthusiasm of motivated researchers can be unbounded when their research is actually helping the company develop new products.

In Maps On Us, we created a very useful Web service. I still use Maps On Us, a couple of times a week on the average, even though it has been several years since we sold it. Every so often, I run into people who use Maps On Us. I feel very proud of having been instrumental in creating it.

Maps On Us gave me the opportunity to learn more about producing and delivering commercial quality software, running a 24 × 7 service, running a venture (even though we never did leave the folds of Lucent), meeting customers, understanding and appreciating sales issues, making a sale, and so on. I now have a very good appreciation of how difficult it can be to make a sale. Researchers are not subject to the tremendous pressure that some of our business colleagues are subjected to. The consequence of not being able to publish yet another paper is unlikely to be as serious as the failure to meet revenue targets. Although I was disappointed that we were unable to spin off Maps On Us as a separate business, the Maps On Us experience made me a more valuable research manager for Lucent.

I was grateful to research chief Netravali for initially putting his trust in me. I got a chance to run a business, even though it was for less than a year. It was unfortunate that I lost Netravali's backing for Maps On Us. Netravali told me recently that he was uncomfortable with my running Maps On Us since I had no prior business experience. Socolof of NVG, a year or so after the sale of Maps On Us, said that I did not properly manage the business expectations with respect to Netravali and NVG.

My lack of business world experience was a drawback, but not one that was insurmountable or one that was unique – many other technical persons who start businesses have to deal with the same issue. And I was learning very quickly. Although I was spending most of my long hours running Maps On Us, I still had some research management responsibilities. Once Maps On Us had been launched, I should have resigned as the head of the Database Systems Research Department and focused solely on Maps On Us. I did not do this because NVG had not put together a plan to spin off Maps On Us with equity for the team members.

The Maps On Us team worked as in a startup – they gave their best. We had never run a 24 × 7 service, but we took on the challenge and delivered.

POSTSCRIPT

About a year and half after we sold Maps On Us to Banyan Systems, there appeared the following headline on the newswires (December 22, 1999):

AOL to Buy MapQuest for $1.1 Billion

MapQuest was one of our competitors. We had done very well in reviews, better than MapQuest in many of them. I sometimes wonder if with more patience Maps On Us could have approached this value.

9 Most Fantastic Place!

T HEORETICAL PHYSICIST PHILIP Platzman, who has been at Bell Labs for over 40 years, worries about the future:

> *I think Bell Labs is the most fantastic place that ever existed. I am very distressed that it's not what it was. People are leaving. ... I don't know who has left in the week I've been away. And I worry that people will not continue to flock to Bell Labs. That would be a loss for the country, and for the world.* [130]

Indeed, Bell Labs, the crown jewel of Lucent, is facing challenging times because Lucent, like its competitors, is facing the pain of a difficult telecommunications market. In this trying period, Bell Labs needs to transform itself into an industrial research lab that pulls Lucent ahead of its competition.

Bell Labs became the greatest research lab of the twentieth century because AT&T was able to pump large amounts of money into it, which allowed Bell Labs to hire the best researchers, buy them the latest equipment, and let them do research unencumbered by business constraints. After 1984, AT&T's revenues were not guaranteed, but they were still substantial enough for AT&T to let Bell Labs operate as before. However, this scenario could not continue for long because AT&T did not fare well in a competitive environment.

Revenues did not grow as AT&T had hoped and, in addition, AT&T was losing market share to its competitors.

Bell Labs needed to change since its parent AT&T had undergone radical surgery. For Bell Labs to operate as before to advance science, Bell Labs had to be funded by an organization, such as the American government, that was not worried about profits and competition. In the monopoly days, the people of the USA were funding Bell Labs by a tax in the form of higher telephone rates. As part of the divestiture, if the American government had wanted Bell Labs to operate as before, Bell Labs could have been funded by the American government, this time from general tax revenues – once again funded by the people of the USA.

Lucent is a much smaller company than the old AT&T that has been forced to shrink dramatically after its birth in tandem with the fall in its revenues. The Bell Labs budget that Lucent can now fund is much smaller because Lucent is operating in an extremely competitive and a very difficult telecommunications market.

RESEARCH PARADIGM

Bell Labs was the crown jewel when it came to advancing science, but not in developing technologies and products for the business units. However, its original mission had been to help the business of its owner AT&T.[131] For example, Bell Labs helped AT&T in providing universal telephone service in the USA. Some time, during the decades of the AT&T monopoly, its mission evolved into being a world-class research institution. According to William O. Baker, the fifth president of Bell Labs, Bell Labs researchers aim to explore nature and science and to determine how these discoveries can be valuable to society.[132]

Business relevance was considered to be a distraction from advancing science. For example, Dennis Ritchie, in his Turing Award Lecture, expressed the following view prevalent in the AT&T monopoly days:[133]

More than anything else, the greatest danger to good computer science research today may be excessive relevance ...

Another danger is that commercial pressure ... will divert the attention of the best thinkers from real innovation to exploitation of the current fad, from prospecting to mining a known lode ...

As long as AT&T was a monopoly, Bell Labs would continue to do basic research and that was fine with every one involved, that is, Bell Labs, AT&T, and the American government. In 1984, the world of Bell Labs' benevolent and generous parent underwent a dramatic change. Bell Labs, which had become adept at basic research, now had to excel at industrial research whose purpose, whether it is short-term or long-term, is to advance the business prospects of the company. In other words, Bell Labs would now have to help its parent make money.

According to Professor Kenneth J. Lipartito, there are two models of innovation in the telecommunication industry.[134] In the first model, the system-oriented research model, research is oriented towards finding incremental improvements to enhance company products and services. Research is structured to serve the various organizations in the company. This model operates when a company is financially comfortable. In such a scenario, the research organization develops a collegial culture with individual researchers having a great deal of freedom to pursue their own research interests.

In the second model, the market-oriented research model, research is focused on the creation of new products and services. The research organization is structured along the lines of product and service groups. This model operates when the company is fighting to grow or even maintain its revenues. In such an environment, the research organization's culture is more corporate than collegial with researchers working on projects that address the competitive needs of the company.

Depending upon the state of the company, its research organization swings between these two models.[135] During periods of competition, when the company has to be aggressive in the marketplace to improve or maintain its business position, the research pendulum swings towards market-oriented research. During periods of stability, when the company is doing fine and can grow revenues and market share or maintain it without having to fight off competition, the research pendulum swings towards system-oriented research.

In extended periods of stability and plentiful resources, there is also a third model, the university-style research model, towards which a company's research gravitates, away from the system-oriented research model, as happened in Bell Labs. Research is focused on advancing science, i.e., basic research, as opposed to developing or improving products and services. Research is organized like university departments and the researchers interact more with university colleagues than with their business colleagues.

Organized research in AT&T and Lucent, with respect to Bell Labs, can be classified into four main periods:

- o 1907–1915: Competitive era for AT&T with research in AT&T following the market-oriented research model.
- o 1915–1984: Monopoly era for AT&T. Bell Labs established 1925. Bell Labs' research model first shifts towards the system-oriented research and then over several decades it shifts to university-style research focusing on basic research.
- o 1984–1995: Competitive era for AT&T and Bell Labs slowly shifts away from university-style research towards market-oriented research.
- o 1996–PRESENT: AT&T hands Bell Labs to its offspring Lucent in 1996. Change in Bell Labs towards market-oriented research continues at a faster pace. Lucent's financial difficulties further accelerate this change.

Graphically, the transition between the various research models before and after the creation of Bell Labs can be represented as

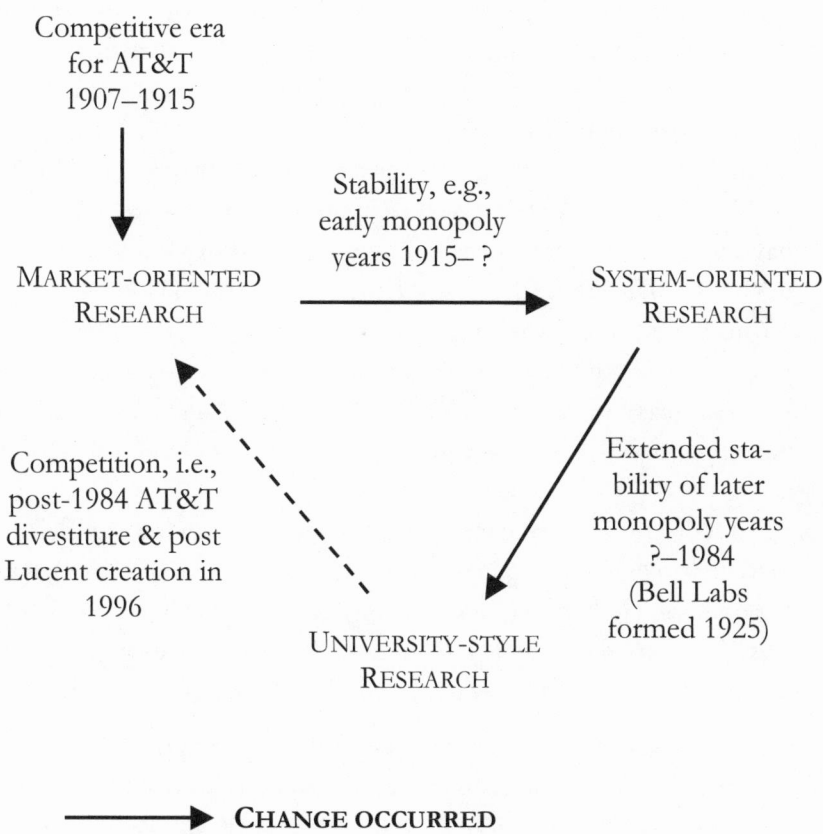

Switching from market-oriented research to system-oriented research and from system-oriented research to university-style research is relatively easy. Researchers are freed from immediate pressures such as product deadlines, they do not have to work under business

constraints or attend numerous meetings, they are free to pick their own research topics, etc.

However, switching from university-style research to market-oriented research is hard because this requires putting constraints on researcher freedoms and requires the researchers to change their behavior. They must now justify the research topics selected by them, their research contributions are now evaluated also for business value in addition to innovation, and they cannot focus on publishing and building their own professional reputations. The researchers now also have to work with business colleagues whose value and reward systems are different from those of the researchers. They need to meet business deadlines, attend numerous meetings, and follow business processes in software development. Projects can be canceled for business reasons. Moreover, in the case of Bell Labs, this means giving up a decades old research culture and working with business units.

According to Dan Stanzione, the eighth president of Bell Labs, market forces have required broad behavioral changes in Bell Labs researchers. He believes that that the behavioral changes in Bell Labs have had far more impact than the technological changes. Perhaps the biggest change that has occurred in Bell Labs has been the

> ... *close working relationships developed across functional lines – for example, between researchers and sales people and extending to customers themselves. Arun Netravali personally led much of this change, as head of Bell Labs Research.* [136]

These behavioral changes are exactly what Bell Labs needs to move away from university-style research towards market-oriented research.

Then there is the issue of working with the business units. When doing basic research, the researchers work on their own, without the involvement of the business units. However, a market-oriented research model necessarily involves working with the business units.

The business units will need to change their behavior and involve Bell Labs in their strategy and product roadmap processes, and this is going to be a tougher task than shifting gears in Bell Labs.

When collaboration between business units and Bell Labs happens willingly, the likelihood of success is very high, for example, consider OCELOT, a software system for optimizing wireless networks:

> ... *deploying a wireless network is a painstaking process. Radio engineers with test gear drive around the intended coverage area, identifying spots where customers may have trouble making calls. ... The OCELOT software virtually eliminates the need for drive tests ... None of our competitors has a tool that can do what the OCELOT software does for service providers and their customers ... And the reason we have it is Bell Labs ...*[137]

The success happened largely because the wireless business unit and the mathematicians and computer scientists from Bell Labs teamed up to work on a problem that was identified by the business unit.

Another successful example of collaboration is Bell Labs and the optical networking business unit (ONG) teaming up to build the LambdaRouter, an optical router that works with tiny mirrors. Bell Labs researchers were able to establish a good relationship with ONG in part because ONG was aware that they did not have the physics expertise to develop some kinds of new optical products. In addition, the physics researchers were happy to work with ONG.

According to the Financial Times, the LambdaRouter is[138]

> ... *the first high-capacity all-optical router. Invented by Bell Labs, the router was introduced commercially earlier this year. ... A rash of new companies want to create all-optical networks that will be faster, more efficient, and cheaper than today's technology. Lucent seems to have stolen a march on these nimble upstarts, however, through Bell Labs.*

The research and business unit teams took about eighteen months to develop the LambdaRouter from concept to product. Two of my software researchers, Bharat Kumar and Dan Lieuwen, also worked on the project. It took proactive action on my part to find that the LambdaRouter team needed help on software development. Kumar and Lieuwen were very excited about working on a project that could help Lucent.

The LambdaRouter was a tremendous technological accomplishment and held good business promise. Unfortunately, it came at just the wrong time. The telecommunications market was in a deep slump because of overcapacity and the service providers had cut down on equipment purchases. The LambdaRouter project was canceled in the fall of 2002.

SHIFTING TO MARKET-ORIENTED RESEARCH

Two key factors for speeding up Bell Labs' shift towards market-oriented research are in place. Lucent's need for innovation is urgent and there is new leadership at Bell Labs.

In the fall of 2001, Netravali resigned as the president of Bell Labs after a short two years on the job:

> [Netravali] has decided to pursue other interests outside of Lucent. Netravali has agreed to work in a new role of Chief Scientist ... he will work with the academic and investment communities to identify new technologies that will be relevant to Lucent's mission, and will act as an advisor to Lucent senior management on technical and customer issues. [139]

Upon his exit in 2001, some researchers breathed a sigh of relief since they held Netravali responsible for "destroying" long-term research. However, Netravali was only trying to do what was needed and expected by the corporation. Netravali's going will not shift the pendulum back to university-style research. In fact, post-Netravali Bell Labs is moving even more quickly and more aggressively towards market-oriented research.

Lucent installed new management at Bell Labs in 2001 to align Bell Labs with Lucent business strategy and marketing efforts.[140] Bill O'Shea, Lucent's executive vice president of strategy and marketing, replaced Netravali as the president of Bell Labs. O'Shea will continue to be Lucent's executive vice president of strategy and marketing, a very important corporate position. Simultaneously with O'Shea's appointment, Jeff Jaffe, who had recently taken over as the head of Bell Labs Research replacing the retiring Bill Brinkman, was appointed as the president of Bell Labs Research and Advanced Technologies.

The new leaders O'Shea and Jaffe, unlike their predecessors Netravali and Brinkman, did not rise up the management ladder through the ranks of Research. O'Shea made his way up the corporate ladder by working in the business units and Jaffe was recruited a few years ago from IBM to head the Advanced Technologies part of Bell Labs. Thus, O'Shea and Jaffe do not have the Bell Labs cultural baggage nor do they have emotional ties to Bell Labs that will prevent them from taking aggressive steps to change research direction and culture.

O'Shea has the credentials to lead Bell Labs into a closer partnership with the business units. He has 30 years experience of working in the AT&T/Lucent business units and has been involved in almost all aspects of business. O'Shea was known as the "boy wonder" in the early 1980s since he had become an executive director at an early age. Wearing two important hats, that of the Bell Labs president and of Lucent's executive vice president of strategy and marketing, gives O'Shea the leverage and clout with both Bell Labs and the business units to ensure that they are on the same page. As Bell Labs president, O'Shea intends to strongly connect Bell Labs Research with the company's business strategy and its customers.[141] O'Shea also plans to ensure that Bell Labs research projects are constantly reviewed to focus research on the areas with the highest pay-off.

O'Shea's deputy, Jaffe, has the operational responsibility to make changes in the Bell Labs research model, ensure that Bell Labs works with the business units, and ensure that Bell Labs delivers a good return on investment to Lucent.

As a first step, by the spring of 2002, Jaffe and his managers had developed a Bell Labs wide strategy document that defined what was important for Bell Labs researchers and managers to work on from a business perspective. All research projects will now have to be aligned with Bell Labs strategy which itself would be aligned with the business units. Execution of the strategy will be key in determining the value of Bell Labs to Lucent.

FUNDING MODEL

Bell Labs is centrally funded with essentially its entire funding coming directly from corporate funds (along with some minor amount of government funding). Many of the big industrial research labs are not 100% centrally funded from corporate funds. For example,[142] research in IBM is funded 65% from corporate funds, 25% by business units, and the rest by royalties and other sources. GE's research is funded 30% from corporate funds, 52 % by business units, and 18% by external contracts. Research in NEC is funded 70% from corporate funds and 30% by the business units. On the other hand, HP Labs and Microsoft Research are 100% centrally funded from corporate funds.

The Bell Labs Research central funding model has been the same for decades. In order to enable researchers to work on topics of their choice without worrying about the business unit pressures or business constraints, AT&T decided to centrally fund Bell Labs from corporate funds. Moving away from 100% central funding, for example, can be one possible step in speeding up the shift to the market-oriented research model. For example, central funding of Bell Labs can be reduced to 75% with the remaining 25% coming directly from the business units and targeted for specific projects. This would allow the business units to have some control on Bell Labs' research direction and ensure that Bell Labs researchers work on projects that help their businesses.

Under such a model, researchers and their managers would have to actively solicit funds from the business units while, at the same time, business units would have to find researchers who will help their business goals. Both Bell Labs and business units would need to be held accountable for ensuring mutual engagement. Such a funding model would lead researchers to work on items relevant to the business units. On the other hand, to ensure cooperation with Bell Labs, the business units will be required to give the 25% to Bell Labs, and to ensure that the business unit managers fund good projects, they should be measured against the return on investment.

To get the 25% funding, Bell Labs researchers and managers will have to work on topics approved a priori by the business units and will have to collaborate with them. To give the money to Bell Labs projects, the business units will have to collaborate with Bell Labs and fund projects that will give them the best return on investment. Such a scenario will force an effective and productive cooperation between Bell Labs and the business units.

THE TWENTY-FIRST CENTURY

Bell Labs is going through difficult times because of the problems of its parent, Lucent, which can no longer fund even the downsized Bell Labs to focus solely on science.

As if Lucent's financial problems were not enough of a headache, Lucent and Bell Labs, facing more cuts, were buffeted by a scandal involving research fraud by Bell Labs scientists, the first such incident in its long history. In 2001, Bell Labs scientists claimed that they had developed a molecular transistor, a nano-transistor, which represented a very significant advance in electronics:

> *Bell Labs, the storied research lab that invented the transistor in 1947, announced the development of a tiny new transistor made of a simple cluster of organic molecules.*

...

Bell Labs' original 1947 transistor measured about an inch in length. Its latest version, whose active portion is just a single molecule across, is about a millionth the size of a grain of sand, ...

— *AP Financial Newswire*, October 18, 2001.

The nano-transistor could lead to tremendous increases in computing power. Most increases in computing power have resulted from reductions in transistor size.

The scientist leading the nano-transistor project seemed to be on a fast track to a Nobel Prize because of this and other extraordinary scientific advances attributed to him. Scientists in other places could not verify these advances, which were in the hot research areas of molecular electronics, molecular crystals, and super conductivity. They could not reproduce the results of the Bell Labs scientists. Rumors about the claims of the Bell Labs scientists being fraudulent and based on cooked up data soon started to surface.

To investigate these rumors, in May 2002, Bell Labs set up an independent committee consisting of five well-known scientists. This committee, which reported its findings on September 24, 2002, found that one Bell Labs scientist had committed fraud and was guilty of misconduct. This scientist had been the lead researcher of the team that had claimed development of the molecular transistor and of several other projects whose results had been published. Bell Labs immediately fired the scientist. Bell Labs also initiated steps to ensure that such research fraud did not occur again.[143, 144]

Moving past this scandal, Bell Labs now needs to aggressively help develop technologies and products for the business units. Some Bell Labs researchers are laying the groundwork for business success with important patented technologies. For example, researchers have developed a wireless communications technology, called BLAST, that uses multi-element antennas at the transmitter and receiver and allows transmission rates far in excess of those possible with conventional techniques. Another example is the Raman amplification technique that allows optical fiber to amplify the signals traveling through it thus allowing signals to travel much farther before they need to be boosted again. One of Lucent's recent and innovative products, the Bell Labs-designed LambdaXtreme Transport, an optical networking system for long distance communication, uses Raman amplification.

MIT's Technology Review magazine voted BLAST one of the five most important patents issued in 2001. The year before, it had voted the Raman Amplification system one of the five most important patents issued in 2000.

A high technology business like Lucent needs Bell Labs to be a technology leader. With Lucent going through a difficult time, skeptics sadly predict the demise of Bell Labs, while Bell Labs leaders continue to say that Bell Labs will continue to operate at the frontiers of research.[145]

Several years ago, Penzias said that Bell Labs had to accept the business reality that it had to help develop technologies for the AT&T business; otherwise, somebody could come along and kill Bell Labs.[146] This still applies today to Bell Labs except that the business reality is now Lucent's business challenges and the difficult scenario in the telecommunications industry. Bell Labs must quickly become, if it can, the industrial research locomotive that, with rapid and timely innovation, pulls Lucent ahead of the competition by giving it an edge with new and innovative products. Researchers will have to compete with the innovation being produced by nimble startups. Product-oriented innovation, provided it is true innovation, can also lead researchers to publications, prizes, and awards!

There are opposing viewpoints on whether or not an in-house research organization can be an effective technology generator in this fast paced world of the Internet, global competition, and focus on core competency. For example, *Business Week* in the article titled "Masters of Innovation" says that, in today's world, in-house R&D is not the right model:[147]

... the new Masters of Innovation must adopt extreme pragmatism, with a ruthless eye on results. Few companies will ever again try to emulate Ma Bell's iron grip on in-house R&D. Today, the best ideas are more likely to be hatched outside and acquired through a partnership or buyout. ... The marketplace, in short, becomes an R&D lab.

On the other hand, companies like Microsoft are slowly building up internal research labs.

Both Bell Labs president Bill O'Shea and Lucent chairman Henry Schacht are convinced that Bell Labs is important to make Lucent a technology leader. O'Shea, speaking at a technology conference in Paris in early 2002, said that he believed that much of the basic research being done at Bell Labs today will bring value to Lucent as the foundation for Lucent products.[148]

Chairman Schacht believes that Bell Labs is the heart and soul of Lucent and will be responsible for providing the technological leadership:

Station: CNBC
Program: Market Wrap
Interviewer: Bill Griffeth, co-anchor
Interviewee: Henry Schacht, Chairman & CEO, Lucent.
Date & Time: July 24, 2001, 4:00–5:00 PM

```
Griffeth: The scouting report on Bell
     Labs is that it's possible that
     through all this turmoil ... Bell
     Laboratories could be behind the
     technology curve and that could be a
     key to your future. What do you say?
```

Schacht: ... I don't think that's right. I think that Bell Labs has been the heart and soul of the place; it's been an innovative engine. And while we will be doing some reduced sizing of Bell Labs, it still remains the heart and soul of the place. ... We mostly think of Bell Labs - us people in the general public, about their research facility. And we will maintain a core of that for our day-after-tomorrow products.

Notes

PREFACE

1. Chairman Henry B. Schacht's message to the Lucent shareowners in the 2000 Lucent Annual Report.

1 I HAVE A JOB FOR LIFE!

2. Lucent Closes Silicon Valley Laboratory. *The New York Times*, page C4, February 7, 2001.

3. David Nagel left AT&T Labs six months later in August 2001 to go to Palm, Inc. A few months later, in January 2002, almost a year after the closing of Bell Labs Research Silicon Valley, AT&T Labs reportedly fired about 40% of its research staff. AT&T Labs was formed as a result of the AT&T trivestiture in 1996 by staffing it with a bunch of Bell Labs researchers.

4. Lucent Technologies says rumors of bankruptcy are baseless. Lucent Press Release, April 4, 2001.

2 THE CROWN JEWEL

5. "Down and Out in Murray Hill" by Irvin Goodwin. *Nature*, vol. 412, August 8, 2001.

6. "Bell Labs: A Bit Abstract and Always Curious" by William J. Broad. *The New York Times*, Section: Business/Financial Desk, May 30, 2001.

7. *Manufacturing the Future: A History of Western Electric* by Stephen B. Adams and Orville R. Butler. Cambridge University Press, 1999, page 115.

8. *Mission Communications* by Prescott C. Mabon. Bell Labs, 1975, page 25.

9. Theodore Vail was the president of AT&T twice, first from 1885 to 1887, and then from 1907 to 1919.

10. See note 8 above.

11. "Strategy and Innovation At Bell Laboratories 1907-1994" by Kenneth J. Lipartito. In *Cnet and its history, Memorandum Excerpt Series (Le Cnet et son histoire, Mémento Hors Série)*, February 1995.

12. *Manufacturing the Future: A History of Western Electric* by Stephen B. Adams and Orville R. Butler. Cambridge University Press, 1999, page 113.

13. Ibid. Pages 113-5.

14. *Mission Communications* by Prescott C. Mabon. Bell Labs, 1975, page 1.

15. By the fall of 2001, Bell Labs consisted primarily of two organizations, namely, Research and Advanced Technologies, each headed by a vice president. Both vice presidents reported to the president of Bell Labs. On October 15, 2001, Jeff Jaffe, vice president of Advanced Technologies, was also appointed as a successor to the retiring vice president of research, Bill Brinkman. At this time, the two vice president positions were consolidated into one new position, the president of Bell Labs Research & Advanced Technologies reporting to the Bell Labs president.

16. Information from the Bell Labs website www.bell-labs.com, 2002.

17. "Nobel Endeavors – An Overview of Nobel Prize-Winning Research at Bell Labs" by Saswato R. Das. In the *Bell Labs Technical Journal*, vol. 5, no. 1, January–March 2000.

18. According to Alfred Nobel's will, prizes are to be given to those who "have conferred the greatest benefit on mankind." The Nobel Prize in Physics is given to the person who is judged in the previous year to "have made the most important discovery or invention within the field of physics." Description taken from the description of Nobel Prizes in Physics at The Nobel e-Museum website www.nobel.se/physics/.

19. The Turing Award, which is awarded by the Association for Computing Machinery (ACM), is the most prestigious award recognizing contributions in computer science and information technology. Many consider it the equivalent of the Nobel Prize for computing. ACM is an educational and scientific computing society.

20. "Scientific Fraud Found at Bell Labs" by Linda A. Johnson, The Associated Press. In the *Seattle Post Intelligencer Online Edition*, September 26, 2002.

21. William B. Shockley's Obituary. *The New York Times*, August 14, 1989.

22. "The Nobel Prize in Physics 1978." www.nobel.se/physics /laureates/1978.

23. "The Nobel Prize in Physics 1997." www.nobel.se/physics /laureates/1997.

24. "The Nobel Prize in Physics 1998." www.nobel.se/physics /laureates/1998.

25. "The Bell Labs – Conditions for Basic Research at Privately Financed Institutions" by Horst L. Störmer. *Innovative Structures in Basic Research* Conference of the Max Planck Society, Ringberg Castle, Germany, October 2000.

26. "An Early Retrospective" by Dennis Ritchie. *Tenth Hawaii International Conference on the System Sciences*, Honolulu, January 1977. A version of this paper is at cm.bell-labs.com/cm/cs/who/dmr/retroindex.html.

27. "UNIX at 25" by Peter H. Salus. *Byte*, October 1994. Also, at www.byte.com/art/9410/sec8/art3.htm.

28. Ibid.

29. David Tilbrook cofounded HCR, the first Canadian UNIX company, and was the program chair of several UNIX conferences.

30. *A Quarter Century of UNIX* by Peter H. Salus. Addison-Wesley, 1994.

31. A recent search for books with UNIX in their titles produced a result of 838 books on www.amazon.com and 1015 books on www.bn.com (Barnes & Noble).

32. *Document Formatting & Typesetting on the UNIX System* by Narain Gehani and *Document Formatting & Typesetting on the UNIX System (Vol. 2)* by Narain Gehani & Steven Lally. They were both published by Silicon Press (www.siliconpress.com) in 1987 and 1988, respectively.

33. See note 30 above.

34. The AT&T logo is sometimes informally referred to as the "Death Star" because of its resemblance to the high-tech spherical fortress called the "Death Star" in the movie "Star Wars."

35. "Bell Labs Fissions, Yielding AT&T Bell Labs and Bellcore." *Physics Today*, May 1984, v37, no. 4, page 77.

36. Ibid.

37. *Three Degrees Above Zero: Bell Labs in the Information Age* by Jeremy Bernstein. Charles Scribner's Sons, 1984, page 214.

38. Before the 1984 AT&T divesture, Bell Labs consisted of about 25,000 employees. After the divestiture, about 3000 employees, including about 100 researchers, went to Bellcore, and 4000 employees went to AT&T Information Services, which had to be kept at arms length from the rest of AT&T according to the 1982 Consent Decree.

39. "Bell Labs Innovations in Recent Decades" by William O. Baker, Ian M. Ross, John S. Mayo, and Daniel C. Stanzione. *Bell Labs Technical Journal*, vol. 5, no. 1, January–March 2000.

40. AT&T spun off NCR six years after acquiring it. The newly freed NCR had a market capitalization of $3.2 billion, which was $4.2 billion less than the $7.4 billion paid by AT&T to acquire it.

41. The process of spinning off Lucent from AT&T was initiated in April 1996 with an initial public offering of Lucent stock. The spinoff was completed on September 30, 1996 when AT&T distributed Lucent shares to AT&T shareholders.

42. Just over two years later, in September 2002, Lucent stock was trading under $1 (after taking into account the Avaya and Agere spinoffs).

43. According to Robert Buderi, writing in the *Engines of Tomorrow*, Simon & Schuster, 2000, central funds represent 97% of Bell Labs Research budget with 3% coming from government contracts.

44. See note 6 above.

45. The National Science Foundation (NSF) uses the following definitions for basic and applied research and development. (a) BASIC RESEARCH: The objective of basic research is to gain a more comprehensive knowledge or understanding of the subject under study, without specific applications in mind. In industry, basic research is defined as research that advances scientific knowledge, but does not have specific immediate commercial objectives, although the research may be in fields of present or potential commercial interest. (b) APPLIED RESEARCH: Applied research is aimed at gaining knowledge or understanding to determine the means by which a specific, recognized need may be met. In industry, applied research includes investigations oriented towards discovering new scientific knowledge that has specific commercial objectives with respect to products, processes, or services. (c) DEVELOPMENT: Development is the systematic use of the knowledge or understanding gained from research directed toward the production of useful materials, devices, systems, or methods, including the design and development of prototypes and processes. These definitions are from the NSF website www.nsf.gov/sbe/srs/seind96/ch4_defn.htm.

46. *Mission Communications* by Prescott C. Mabon. Bell Labs, 1975, page 94.

47. Of course, a researcher who had a good year would be treated better, say from a salary perspective, than a researcher who had a bad year.

48. See note 25 above.

49. "Business unit" is the term used for a somewhat independent business organization, a profit center focusing on selling related products and services. The number and the area of responsibilities of the business units were not fixed

and these changed over time. For example, Lucent now has only two business units, Integrated Network Business Solutions and Mobility Solutions.

3 LIFE AT MURRAY HILL

50. A few months before my transfer, the PWB/UNIX development organization, which I had joined, was reorganized. As a result, I became part of an advanced development department in Piscataway, which was relocated to Murray Hill while I was in the process of transferring to Research.

51. See note 30 above.

52. *Mission Communications* by Prescott C. Mabon. Bell Labs, 1975, page 4.

53. Ibid. Pages 5 and 6.

54. *Engines of Tomorrow* by Robert Buderi. Simon & Schuster, 2000, page 18.

55. "Concurrent C" by N. Gehani and W. D. Roome. *Software Practice & Experience*, vol. 16, no. 9, September 1986.

56. "Ode: Object Database & Environment" by R. Agrawal and N. Gehani. *SIGMOD*, Portland, Oregon, 1989.

57. *Three Degrees Above Zero: Bell Labs in the Information Age* by Jeremy Bernstein. Charles Scribner's Sons, 1984, page 230.

58. Jeff Jaffe, vice president of Advanced Technologies, also became vice president of research on October 1, 2001. These two positions were then combined into the new position of president of Bell Labs Research & Advanced Technologies. Advanced Technologies is the Lucent internal consulting organization that does contract work for the Lucent business units.

59. "Bell Labs Chief Laments 'Fragile' State of Research" by Terry Costlow. *Electronic Engineering Times*, January 1, 2000.

60. *Engines of Tomorrow* by Robert Buderi. Simon & Schuster, 2000, pages 26 and 27.

61. See note 55 above.

62. See note 56 above.

63. Towards the end of 2001, the Database Systems Research Department was renamed as the Network Data and Services Research Department reflecting the changing needs of Lucent.

64. Centers are Bell Labs Research organization units that are headed by research vice presidents. Each center consists of several departments. A center in Bell Labs used to have between 75 to 100 researchers, but now the number has shrunk to about 50 or so because of downsizing Bell Labs and Lucent spinoffs and financial problems.

65. Ken Thompson was awarded an honorary PhD (Doctor of Science, to be precise) by McGill University, Montreal, Canada in recognition of his contributions in conceiving of and developing the UNIX system.

66. *Engines of Tomorrow* by Robert Buderi. Simon & Schuster, 2000, page 264.

67. The review process for the managers was similar to that of the researchers. For example, in case of the directors, a committee consisting of their vice presidents and their senior vice president evaluate the directors' performance.

68. Softswitch is a new kind of switch, cheaper and more flexible than a traditional telephone switch, which operates with both Internet and traditional telephone networks.

69. See note 66 above.

70. Bell Labs organization numbers change occasionally. The numbers given in this book were the numbers in use when I left Bell Labs in 2001.

71. When I joined Bell Labs in 1978, Bell Labs had development organizations that developed products for the business parts of AT&T. Several years later, following the 1984 AT&T divestiture, they were "moved" out of Bell Labs and put under the purview of the business units for whom they were building products.

72. "A Comparison of the Programming Languages C and Pascal" by A. Feuer and N. Gehani. *ACM Computing Surveys*, vol. 14, no. 1, March 1982.

73. *The C Programming Language (2/E)* by Brian W. Kernighan and Dennis M. Ritchie. Prentice Hall, 1989.

74. "Why Pascal is Not My Favorite Programming Language" by Brian W. Kernighan. In *Comparing and Assessing Programming Languages*, edited by Alan Feuer and Narain Gehani, Prentice-Hall, 1984.

75. *C: An Advanced Introduction* by Narain Gehani. Computer Science Press, 1985.

76. Approval from the Book Review Board was required even though I wrote the book outside of normal work hours during early mornings, late nights, and on weekends. I had to get the Board's approval since I used the Bell Labs computer to write the book and the typesetter to print a camera-ready copy. Moreover, the book was on a topic of interest to Bell Labs and my Bell Labs affiliation was going to be listed on the book.

77. See note 55 above.

4 LOOKING FOR DUNG BUT FINDING GOLD

78. "The Cosmic Microwave Background Radiation," the *1987 Nobel Lecture,* by Robert Woodrow Wilson. www.nobel.se/physics/laureates/1978/wilson-lecture.pdf.

79. "Penzias and Wilson's Discovery is One of the Century's Key Advances." www.bell-labs.com/project/feature/archives/cosmology.

80. SIGMOD, VLDB, and ICDE are annual events, while EDBT is held every two years. SIGMOD is the Special Interest Group on Management of Data Conference, VLDB is the International Conference on Very Large Databases, ICDE is the International Conference on Data Engineering, and EDBT is the Conference on Extending Database Technology.

81. Even if the names of the authors of a paper are hidden from a reviewer, it is often not too difficult for a reviewer to deduce the names of one or more of the paper's authors.

82. Many decades earlier, AT&T had been a global corporation. However, in 1925, Walter Gifford, the new AT&T president, decided that the Bell System should focus on providing universal telephone service within the USA. He withdrew AT&T from the international arena by selling International Western Electric to ITT. Following this sale, AT&T retained an international manufacturing presence only in Canada.

5   DO WE WORK FOR THE SAME COMPANY?

83. Nationwide, AT&T employed about a million people in the 1980s before its divestiture in 1984.

84. The divestiture also allowed AT&T to enter the computer business and other companies to enter AT&T's once protected and very lucrative long distance business. AT&T, as part of the divestiture agreement, spun off its local calling business by creating the Regional Bell Operating Companies, which were also known as the Baby Bells or RBOCs.

85. Lucent acquired one company in 1996, three companies in 1997, eleven in 1998, fourteen in 1999, nine in 2000, and zero in 2001 and 2002. Towards the end of 2000 and in 2001 and 2002, Lucent was doing the opposite, spinning off companies, and selling divisions.

86. "Reflections on Software Research," the *1983 Turing Award Lecture,* by Dennis M. Ritchie. *CACM*, vol. 27, no. 8, August 1984.

87. The optical networking group is part of Lucent's Integrated Network Business Solutions (INS) business unit.

88. When AT&T split up again in 1996, the Business Communications Services (BCS) business unit stayed with AT&T while Bell Labs was handed to Lucent. Lucent renamed its enterprise business unit as Business Communications Systems, which was also known as BCS. The Lucent BCS was spun out in 2000 as Avaya.

89. AT&T acquired NCR in 1991 and restored its independence by spinning off NCR in 1997 as part of the AT&T trivestiture, which also launched Lucent as a separate company.

90. See note 88 above.

6 WHAT ARE YOU DOING FOR US?

91. Bill Brinkman's formal title at the time was executive director of the Physics Research Division.

92. *Engines of Tomorrow* by Robert Buderi. Simon & Schuster, 2000, pages 37 and 38.

93. I say product quality as opposed to full-fledged products because the systems were lacking features such as sophisticated graphical user interfaces (GUIs) and user documentation that come with products.

94. *Engines of Tomorrow* by Robert Buderi. Simon & Schuster, 2000, pages 266 and 267.

95. "Netravali Replaces Penzias as Bell Labs Research VP." AT&T Press Release, October 16, 1995.

96. "Arun Netravali Named President of Bell Labs." Lucent Press Release, October 26, 1999.

97. "Bell Labs Experts Foresee Radical Changes in Communications in the Third Millennium." Lucent Press Release, November 12, 1999.

98. "Bell Engineers: Calm Amid Lucent's Storm" by Terry Costlow. *Electronic Engineering Times*, May 28, 2001, page 133.

99. See note 59 above.

100. *Engines of Tomorrow* by Robert Buderi. Simon & Schuster, 2000, page 171.

101. In spring 2001, Jeff Jaffe, in addition to being appointed the research senior vice president, was also the vice president of Advanced Technologies. Jaffe is now the president of Bell Labs Research & Advanced Technologies and reports to the president of Bell Labs.

102. The Turing Award is considered by many to be the equivalent of Nobel Prize for computing.

103. The lifetime employment culture of Bell Labs started changing in 2001 because of Lucent's financial misfortunes. For the first time, Bell Labs researchers were "terminated involuntarily" (fired) based on criteria such as location (several remote locations were closed) because of the need to reduce the number of employees.

104. "Masters of Innovation" by Amy Cortese. *The Business Week 50*, Bonus Issue, Spring 2001.

105. A letter by Arun Netravali. *Reader's Report, Business Week,* May 7, 2001.

106. The DNS business unit was later folded into a new Lucent business unit, the InterNetworking Systems (INS) business unit.

107. Bill O'Shea is now Lucent's executive vice president of Strategy and Marketing and president of Bell Labs.

108. "Lucent Technologies Introduces Revolutionary IP Switch ..." Lucent Press Release, May 27, 1998.

109. Carly Fiorina is now the chairman & CEO of Hewlett-Packard.

110. See note 108 above.

111. "Lucent Technologies to Acquire Nexabit Networks." Lucent Press Release, June 25, 1999.

112. The term "Class 5" switch is AT&T's terminology for the type of switch that is used in a local telephone company central office.

113. See note 108 above.

114. Competitive local exchange carriers.

115. Incumbent local exchange carriers.

116. "Nortel Playing Hard Ball on the Softswitch" by Scott Moritz. The Street.com, November 5, 2001.

117. "Lucent, Level 3 Partner to Make 'Softswitch' Voice Technology the Foundation of Next Generation Broadband Networks ..." Lucent Press Release, June 23, 1999.

118. "Go West" by Henry Goldblatt. *Business 2.0,* November 1999.

119. "Lucent Technologies to Acquire Excel Switching." Lucent Press Release, August 18, 1999.

120. "Lucent Technologies Introduces Second-Generation Softswitch." Lucent Press Release, January 16, 2001.

121. BCS was spun out of Lucent as Avaya in 2000.

122. Ravi Sethi is now the president of Avaya Labs.

123. The Research/BCS collaboration was initially named DLS, initials that ostensibly stood for the project name "**D**efinity™ / Intuity™ Audix™ **L**DAP **S**ystem", but informally they stood for "**D**on't **L**ook **S**tupid."

7   BELL LABS GOES WEST

124. "Lucent Technologies Opens Bell Labs Research Organization in Silicon Valley." Lucent Press Release, July 1, 1998.

125. Organizationally, BLRSV was a part of Bell Labs Research. But BLRSV was setup with very different goals from the rest of Bell Labs Research. Therefore, in this chapter, when I talk about Bell Labs Research (or just Research), I mean all of Bell Labs Research except BLRSV.

126. Bell Labs Research did hire some research staff without PhDs, but they were primarily software developers, not researchers. The number of software developers in Research was small because Research management had in the past also expected researchers to build software systems, especially prototypes.

127. The world of Wall Street investment bankers, analysts, etc., is very different from that of Bell Labs. However, comparing Wall Street and Research compensation is appropriate because many Murray Hill researchers live in the same communities as the investment bankers and analysts. Wall Street compensation is strongly tied to performance with excellent performers getting huge bonuses. Conse-

quently, many investment bankers, analysts, etc., work very hard, putting in long hours. Despite the lack of such financial incentives at Bell Labs in the past, many researchers worked very hard and long hours because they were passionate about their research. However, the situation is different now. Bell Labs researchers who help the company develop products now stand to get large bonuses and options.

8   MAPS ON US

128. Months later, Mel Cohen was gracious enough to admit that he was wrong in not initially appreciating the value of Maps On Us. Cohen said that with the increasing use and visibility of Maps On Us, I was ensuring that he was constantly reminded of his initial lack of enthusiasm for Maps On Us.

129. Some years later, Sethi praised me for my focus and determination in moving Maps On Us forward.

9   MOST FANTASTIC PLACE!

130. "Bell Labs Research Regroups as Parent Lucent Shrinks" by Toni Feder. *Physics Today*, October 2001, vol. 54, no. 10.

131. To be precise, Bell Labs' owners were AT&T and Western Electric, but AT&T owned Western Electric.

132. *Mission Communications* by Prescott C. Mabon. Bell Labs, 1975, pages 5 and 6.

133. See note 86 above.

134. See note 9 above.

135. See note 9 above.

136. See note 39 above.

137. "Lucent Technologies Software Improves Capacity and Coverage of Wireless Networks, Reduces Deployment Time and Cost." Lucent Press Release, October 11, 2000.

138. "LambdaRouter Benefits Users." *Financial Times*, September 19, 2001.

139. "Lucent Technologies Appoints Bill O'Shea President of Bell Labs, Succeeding Arun Netravali." Lucent Press Release, October 15, 2001.

140. "New Bell Leadership May Help Lucent." *AP Newswire*, October 15. 2001.

141. "Bell Labs gets Lucent Exec as New Chief." *The Star-Ledger*, October 16, 2001, page 21.

142. *Engines of Tomorrow* by Robert Buderi, Simon & Schuster, 2000.

143. "Bell Labs Fires Star Researcher" by Linda A. Johnson, The Associated Press. In the *Chicago Tribune Online Edition*, September 25, 2002.

144. "Panel Says Bell Labs Scientist Faked Discoveries in Physics" by Kenneth Chang. *The New York Times*, September 25, 2002, page A1.

145. "Bell Labs Research Regroups as Parent Lucent Shrinks." *Physics Today*, October 2001, vol. 54, no. 10.

146. *Engines of Tomorrow* by Robert Buderi, Simon & Schuster, 2000, page 281.

147. See note 104 above.

148. Bill O'Shea in a speech at a technology conference sponsored by *Les Echos*, Paris, February 14, 2002.

Index

C

N

O

P